W0017024

THE WONDERS OF LIGHT

Discover the spectacular power of light with this visually stunning celebration of the multitude of ways in which light-based technologies are shaping our society.

Be inspired by state-of-the-art science
Sixteen beautiful, straightforward chapters demonstrate the science behind the fascinating and surprising ways in which light can be harnessed and used, from displays, solar cells, and the internet to advanced quantum technologies.

Be dazzled by brilliant color
Dramatic design and radiant color illustrations bring cutting-edge science and ground-breaking innovations to life, clearly explaining the fundamental principles behind them.

Be part of something bigger
Featuring a foreword by Nobel Laureate and former US Secretary of Energy, Steven Chu, this book has been published in association with ICFO - The Institute of Photonic Sciences to celebrate the 2015 International Year of Light. It will enthrall anyone interested in the developments of science, technology and human civilization that have been made possible by light.

Marta García-Matos is a physicist, mathematician and writer working at the outreach division of ICFO - The Institute of Photonic Sciences.

Lluís Torner is the Founding Director of ICFO - The Institute of Photonic Sciences. He is a Fellow of the European Physical Society, the European Optical Society, and the Optical Society of America.

CAMBRIDGE
UNIVERSITY PRESS

University Printing House, Cambridge CB2 8BS, United Kingdom

Cambridge University Press is part of the University of Cambridge.

It furthers the University's mission by disseminating knowledge in the pursuit of education, learning and research at the highest international levels of excellence.

www.cambridge.org
Information on this title: www.cambridge.org/9781107477414

© Institute of Photonic Sciences (ICFO) 2015

Graphic design: bpdisseny – Anna Jordà

This publication is in copyright. Subject to statutory exception and to the provisions of relevant collective licensing agreements, no reproduction of any part may take place without the written permission of Cambridge University Press.

First published 2015

Printed in the United Kingdom by Bell and Bain Ltd

A catalog record for this publication is available from the British Library

ISBN 978-1-107-47741-4 Paperback

Cambridge University Press has no responsibility for the persistence or accuracy of URLs for external or third-party internet websites referred to in this publication, and does not guarantee that any content on such websites is, or will remain, accurate or appropriate.

THE WONDERS OF LIGHT

MARTA GARCÍA-MATOS
Institute of Photonic Sciences (ICFO)

LLUÍS TORNER
Institute of Photonic Sciences (ICFO)

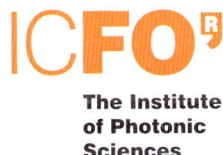

**The Institute
of Photonic
Sciences**

CONTENTS

FOREWORD

Steven Chu
Stanford University
Nobel Laureate
Former US Secretary of Energy (2009–2013)

We are visual beings, and qualities such as "insight" and "vision" describe our understanding well beyond sensory inputs. The ability to extend our sight beyond our eyes gives deeper meaning to the observation of Yogi Berra, the great American philosopher of the 20th century: "You see a lot by just watching."

Visible light is a small sliver of a huge spectrum that includes radio waves with energies as low as 2×10^{-13} eV (48 Hz) used to communicate with submarines, to cosmic gamma rays with energies in excess of 300 GeV (7×10^{25} Hz). Maxwell's equations, written in 1862, predicted that radio waves, infrared radiation, and visible light are different forms of electromagnetic waves. In the ensuing 150 years, we discovered X-rays and gamma rays, and subsequently realized that these forms of energy are also part of the electromagnetic spectrum.

With the development of Quantum Mechanics in the 1920s, we discovered that the energy of these waves cannot be dialed down arbitrarily, but there exists a fundamental "graininess" to light: the photon. An electromagnetic wave of frequency ν is composed of particles of light with energy $E_{min} = \hbar\nu$, where Planck's constant h is a universal constant of Nature. Light can display both particle and wave properties. Remarkably, Maxwell's equations predicted that photons with energies less than 10^{-13} eV to greater than $3 \times 10^{+11}$ eV all move at the same speed in vacuum. This prediction, yet to be contradicted by experiment, is one of the wonders of light.

Marta García-Matos' and Lluís Torner's *Wonders of Light* presents a delightful smorgasbord that illustrates how light continues to redefine our daily lives. The chapter "Lighting" reminds us how artificial light liberated us from darkness (and boredom) and protected us from predators; rapidly advancing technology is allowing us to create light that mimics the changing hues of sunlight during the day. The chapter "Displays" shows how light enhances our ability to visualize. All of these technologies rest on recent stunning advances in the chapter "New Materials."

Our ability to image the microscopic world down to the atomic level helps us understand and combat "Virus Attacks." The chapter "Focus" describes recent advances in the effective resolution of optical microscopy that are able to circumvent the limits imposed by the Uncertainty Principle of Quantum Mechanics. "Optogenetics" speaks of powerful new ways of stimulating neurons in fully functioning animals that are leading to a deeper understanding of the brain, and will soon find applications in pain control and neuronal disorders.

Part of the wonder of light is that it can be used in applications that demand wildly different properties. Light can be extremely "Sharp" for precise laser surgery and destruction of selected cells. It can be "Gentle," and used to cradle and move individual cells and biomolecules. While intense beams of light can heat matter to temperatures that mimic the center of the Sun, it can be used to create "Cold" more than a billion times colder than the temperature of the most remote corners of our universe. Lasers can be utilized in seemingly diametrically opposite applications. They can be ultra-stable clocks that detect a deviation of less than one second over the lifetime of the universe or, alternatively, be used to take ultra-"Fast" snapshots of changes in the electrons of atoms, molecules, and materials. During this time, the atomic motion appears frozen. While a laser beam is synonymous with a straight-line path, and more generally the path that minimizes the transit time between two points in space and time, the chapter "Random Walks" introduces the reader to the manner in which we exploit light when it travels in media that scatter and absorb the photons.

Advances in science and technology do not require an intuitive understanding of the underlying phenomena. The chapters "Privacy" and "Riddles" discuss applications of quantum states connected to each other in ways far removed from human experience. Nevertheless, we used Quantum Mechanics to invent the transistor, the laser, and more recently, quantum computing and quantum cryptography. Over the past half a century, we have learned to live in peaceful co-existence with a wildly successful theory that still defies human intuition. "Connected" shows how we use light to transmit information at staggering rates. While we routinely communicate with people half-way around the world in a blink of an eye, we have no mechanical insight as to how light travels in vacuum. All that remains of the mechanical model Maxwell used to arrive at his equations are the equations: like the Cheshire Cat in *Alice in Wonderland*, the body of the cat disappears and all that remains is its smile. Massive particles, electrons, atoms, molecules also have wave-like properties, but today there is a debate as to whether the quantum phase frequency ν, given by $E = mc^2 = h\nu$, that appears in our most general theory of Quantum Mechanics is measurable in principle.

The Wonders of Light also discusses how light will help us transition to a sustainable world. Because of our ability to exploit fossil energy, a growing number of people keep their homes warm in the winter, cool in the summer, and lit at night. Many of us go to the local market in cars with the power of over a hundred horses and fly across continents in wide-body airplanes with the power of a hundred thousand horses. The Sun is ultimately the source of this energy. What took Nature hundreds of millions of years to create in coal, oil, and natural gas, we are using up in hundreds of years.

Long before we run out of fossil fuels, there is another danger. The carbon emissions used to generate the equivalent of more than a billion horses working continuously have created significant climate-change risks. The overwhelming consensus among climate scientists is that the Earth is warming up due to the burning of fossil fuels, and we need to drastically reduce carbon emissions by mid-century. To accomplish this task, we must convert massive amounts of sunlight energy into electricity. The chapter "Catch that Energy!" describes how the technology is racing ahead. A wonder and necessity of light is its role in providing a clean source of energy necessary to protect us from dire peril.

Silvia Carrasco
Maria Rosa Birulés
Lydia Sanmartí-Vila
John Calsamiglia
Mariona Costa
Pitusa Matos
Jordi Andilla
Jonatan Brask
Juli Céspedes
Tomás Charles
Brook Hardwick

ACKNOWLEDGMENTS

We are greatly indebted to many colleagues and friends who have contributed their generous efforts and invaluable advice to this book, including:

Alina Hirschmann
Marina Mariano
Míriam Martí
Alejandro Sáiz
Susana Santos
Roger Tudó
Chaitanya K. Suddapalli
Albert Verdeny
Benjamin Wolter

As well as, of course,

Steven Chu

All Advisors listed at the end of each chapter

All ICFOnians

ABOUT THE BOOK

"The sun shone, having no alternative, on the nothing new." Samuel Beckett's acclaimed opening words for *Murphy* appear to be somehow pessimistic. Yet actually, quite often, a distinctive control over light beams puts forward a path to the new. This book takes a glance over some of those paths.

Exploring the natural phenomena around us and across the Universe; pushing the limits of our understanding of Nature; and using the knowledge acquired to diagnose and cure disease, and to create devices and machines that make life safer, healthier, and more fulfilling is a program that many, throughout history, have found as captivating as compelling. Advancing such programs requires tools, which must be more and more sophisticated as the limits of knowledge move farther from our own scale and intuition. Really small or really large things, as well as really fast or really slow events, are particularly challenging, as neither our senses nor our ordinary gadgets are equipped to tackle them. We need powerful tools and technologies to enter such territories. The farther we aim to reach, the higher the performances that our toolkit must deliver.

Light is one of these wonderful tools. It is ubiquitous and universal, and can be outstandingly accurate and precise. During the past decades we have learnt not only how to generate and control light in exquisite ways – especially since the invention of the laser half a century ago – but also how to transmit it and display it in ways that used to be the realm of science fiction and novels.

As a result, light-based technologies are, literally, everywhere. And what is available today is just the beginning.

The book consists of sixteen chapters of the same length and structure, each addressing a particular scientific and technological challenge in which some of the multifaceted existing light–matter interactions take a leading role. Readers can go through the chapters following any order they want. Each chapter opens with a short story that aims at motivating a context in which the overall challenge to be addressed has an impact. A brief description of the science and technology of light involved in the solution follows, supported by a set of graphic illustrations and complemented with a technical glossary and suggestions for further reading. Chapters are intended to be self-contained.

We avoided an introductory chapter reviewing the properties of light. Instead, we chose to spread the technical and scientific content through the different chapters, picking up properties only as they were relevant for the different applications.

The first part of the book addresses a wealth of accomplishments that require the finest control of light in order to push fundamental limits of experimental science: achieving the coldest temperatures, just billionths of billionths of a degree over absolute zero; resolving the smallest structural details, at the scale of a billionth of a meter; or filming the fastest processes ever recorded, such as the electronic transitions in chemical reactions that take millionths of billionths of a second.

The second part deals with understanding technologies and applications that have a central impact in our life style and in the way current civilization is shaped. We address applications in health and life sciences, global communications, new materials, renewable energy, lighting sources, consumer gadgets, and futuristic technologies, such as quantum computing. Unfortunately, we had to omit many other applications, in the hope that the material presented here motivates readers to search for more.

BIG BANG!

Since the origins of the Universe...

...and the formation of the Solar System...

...the nature of light...

$$\alpha = \frac{1}{4\pi\,\varepsilon_0}\,\frac{e^2}{\hbar c}$$

$H_2O + CO_2 + LIGHT$

$ORGANIC\ MATTER + O_2$

The science and technology of light has reached a point in which, under the proper conditions, it can be used to direct the exact energy and information we want to the exact point in time and space we need. This is the principle behind an emergent technology, often known as photonics, already shining in a wealth of ordinary situations, but which promises to get in through every and many more cracks in the years to come.

TO FUTURE

...though described as a unified phenomenon...

...holds in limitless wonders...

...of unbounded transformative power.

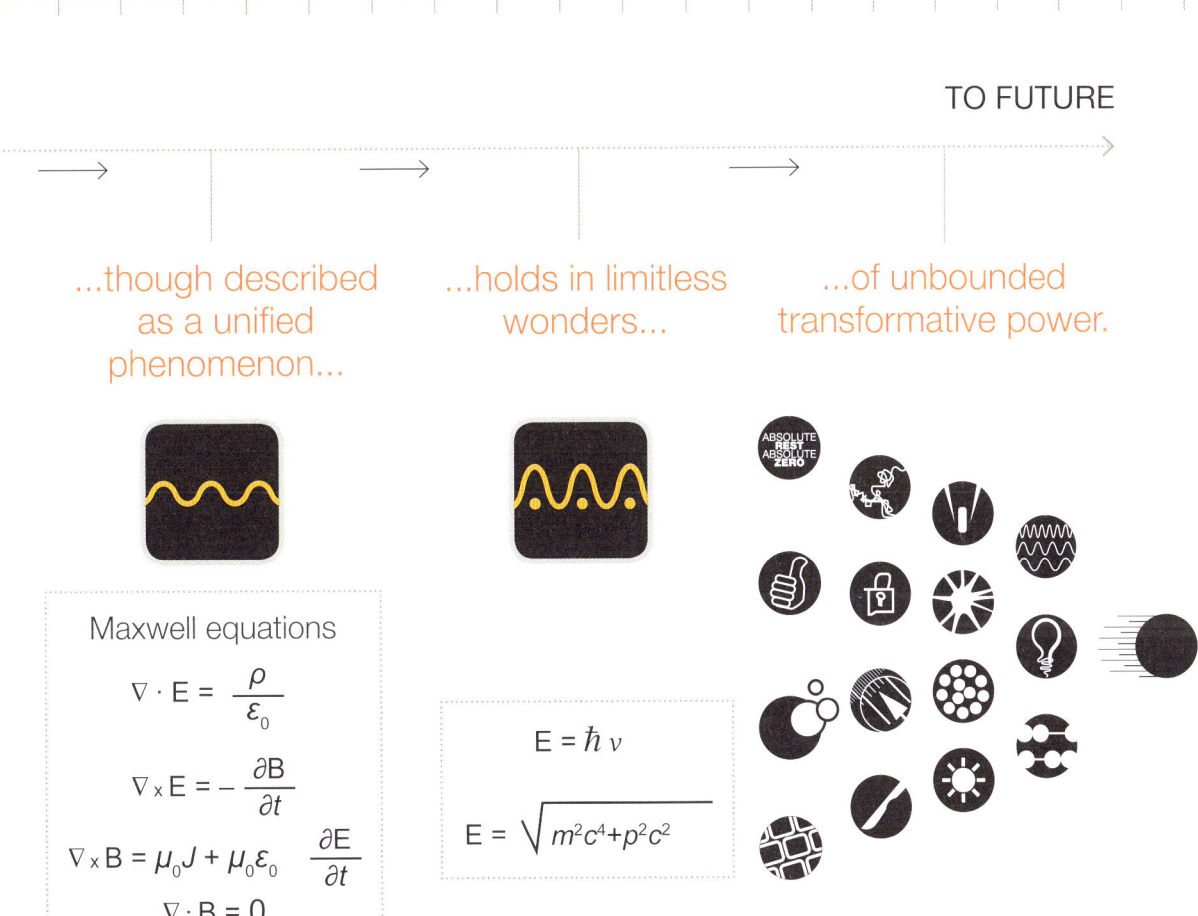

Maxwell equations

$$\nabla \cdot E = \frac{\rho}{\varepsilon_0}$$

$$\nabla \times E = -\frac{\partial B}{\partial t}$$

$$\nabla \times B = \mu_0 J + \mu_0 \varepsilon_0 \frac{\partial E}{\partial t}$$

$$\nabla \cdot B = 0$$

$$c = \frac{1}{\sqrt{\mu_0 \varepsilon_0}}$$

$$E = \hbar \nu$$

$$E = \sqrt{m^2 c^4 + p^2 c^2}$$

COLD

Cold, calculating, and extremely sensitive

COLD

In the nineteenth century it was discovered that matter could not get colder than −273.15 degrees in the Celsius scale. A new temperature scale named Kelvin – after physicist William Thomson, Lord Kelvin – renumbered Celsius to assign zero to that newfound lower bound. Zero degrees Kelvin corresponds to the absolute zero of temperature, the absolute cold.

At the absolute zero all motion ceases to exist. The random movement of the microscopic constituents of matter ceases, and every degree of freedom remains frozen and under control. This is certainly not a pleasant place to be, but those are the idyllic conditions for errorless machine performance, in which computers, precision instruments, compasses, and diagnosis tools will work unaffected by the detrimental effects of thermal noise.

As appealing as it seems, the absolute zero is unreachable. Even in deep space, the background radiation filling the whole Universe since the Big Bang "keeps the vibe" at 3 K. If you were lucky enough to get to some interstellar objects made of expanding gases, like the Boomerang nebula, you could get cooler than that and drop to 1 K. Indeed, cooling by expansion of gases is the principle behind freezers and air conditioning, and it also explains why spray deodorants are so cold. The principle was discovered (again) by Lord Kelvin and had a central role in the development of thermodynamics, the physics of heat and cold. Learning to direct heat pushed steam trains forward while schemes to procure cold revolutionized the industry and commerce of food and products. At a more fundamental level, it propelled low-temperature labs worldwide into a hectic race to get closer and closer to the unreachable absolute zero. This race holds the key to understanding why we are now reaching temperatures of only millionths of billionths of a degree (and dropping).[1]

In 1908, the physicist Heike Kamerlingh Onnes got himself a prominent place in the competition, achieving a record temperature of 2.4 K with the liquefaction of helium. However, Onnes was not interested in the record itself, rather his motivation was confirming the existing theories about the behavior of metals at very low temperatures. He did not, however, find what he expected. Instead, he observed the first macroscopic manifestation of quantum mechanics. By then it was 1911, and the foundations of quantum mechanics were being laid. It took several years to understand that Onnes' surprising experiments produced an exotic state of matter in which electrical resistance vanishes: superconductivity.[2] Nowadays, superconductors support super-fast levitating trains and yield high-definition brain images.

At very low temperatures, quantum phenomena find a peaceful environment in which to emerge. Cold reduces almost all noise and uncontrolled mechanisms in atoms, revealing their quantum-mechanical behavior at large scales. If we could manipulate the quantum information stored in atoms, we could perform quantum computation, metrology, and sensing tasks that outperform their classical counterparts. This is why we want it cold. Light, through laser cooling, is the number one cooling agent, and through light–matter interactions it gives us direct access to and control of the quantum states of atoms.

---

[1] Yes, probably, the coldest points in the whole Universe are the low-temperature labs!

[2] The microscopic theory of superconductivity at low *T* was provided by Bardeen, Cooper and Schrieffer in 1957. We are still missing a fully satisfactory explanation of high-temperature superconductivity.

High temperature

means intense agitation and collisions.

At **low** temperatures, particles are stiller, more calm and ordered.

At temperatures near the

absolute zero

atoms get **ultracold, calculating… and very, very sensitive.**

We could get extraordinary things from them, but they need to be treated with the lightest touch.

Laser cooling is a key step in the quest for **absolute zero**. It allows us to record microkelvin and subsequently nano- and picokelvin temperatures. At these regimes, quantum light–matter interactions find applications in quantum technologies as well as in the study of fundamental physics.

The mechanism behind laser cooling arises from the quantum features of quantum light–matter interaction. Photons are energy packages. Electrons, orbiting the atoms structured in **quantized** energy levels, can absorb or emit a photon when the energy it carries matches the energy difference between electronic transitions. Laser-cooling techniques enforce that when light is sent through an atomic ensemble, each atom is cooled by absorbing and re-emitting photons.

When the photons leave the system, they take with them most of the heat and leave the atoms almost at rest – almost at absolute zero.

In a standard laser-cooling setting, a sample of alkali atoms (sodium, potassium, rubidium, cesium) is cooled to around 100 microkelvin with a combination of magnetic and optical forces. **Near-infrared** laser beams of a few milliwatts in all six directions slow down the atoms as a result of photon–atom **momentum** transfer, cooling the sample below 100 microkelvin in a few milliseconds. Evaporative cooling to a few nanokelvin is then achieved by confining the atoms in a **magnetic trap** and allowing the hottest atoms to escape. Further cooling to picokelvin is demonstrated by removing more energy from the atoms interacting with their **internal spin state**.

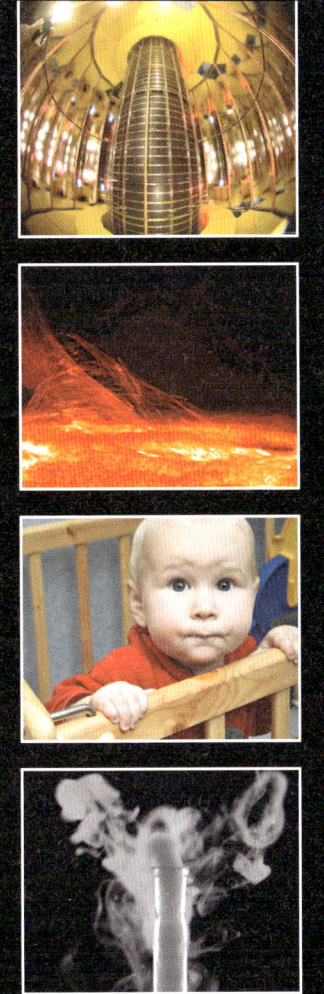

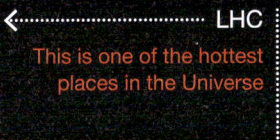

LHC | 10^{16} K

This is one of the hottest places in the Universe

SUPERNOVA EXPLOSION | 10^{11} K

SUN | CORE: 10^7 K SURFACE: 6000 K

HOTTEST PLACE ON EARTH | 330 K

HUMAN BODY | 310 K
This is us

COLDEST PLACE ON EARTH | 183 K

LIQUID NITROGEN | 77 K

DEEP SPACE

This is as cold as it gets out there. Going under takes human-like technology | 3 K

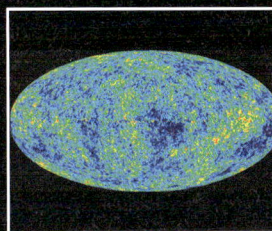

HELIUM GOES SUPERFLUID | 2.2 K

LASER-COOLED ATOMS | 10^{-5} K

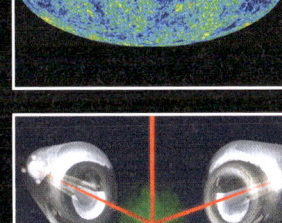

BOSE–EINSTEIN CONDENSATE | 10^{-9} K

This is one of the gates to the coldest places in the Universe | 10^{-12} K

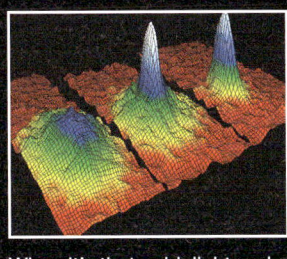

When it's that cold, light and matter interact in smooth ways, and the whole light–matter interface performs as an information long-term storage system or an ultraprecise measuring device.

ABSOLUTE **REST** ABSOLUTE **ZERO**

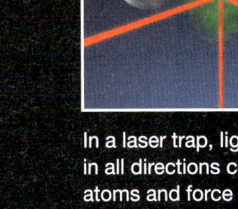

In a laser trap, light beams in all directions collide with atoms and force them to almost a standstill, lowering their temperature to almost absolute zero.

Scientific and technical advisors

MORGAN W. MITCHELL
ICREA professor at ICFO

JORDI MIRALDA
ICREA professor at the Institute of
Cosmos Sciences, University of
Barcelona

LETICIA TARRUELL
Professor at ICFO

Glossary

ABSOLUTE ZERO The lowest possible temperature that is theoretically possible, at which the motion of particles that constitutes heat ceases completely. It corresponds to $-273.15\ °C$ on the Celsius temperature scale and to $-459.67\ °F$ on the Fahrenheit temperature scale.

INFRARED The portion of the electromagnetic spectrum extending from the red end of the visible region to the microwave range. The infrared range is usually divided into three regions: near-infrared, from 700 nm to 2 μm; medium-infrared, from 2 μm to 4 μm; and far-infrared, from 4 μm to 1 mm.

LASER Acronym for Light Amplification by Stimulated Emission of Radiation. A laser produces a highly coherent, highly directional, and nearly monochromatic beam of light, by stimulating atoms or molecules to emit light at particular wavelengths and amplifies that light, typically producing a very narrow beam of radiation.

MAGNETIC TRAP An apparatus which uses a magnetic field gradient to trap neutral particles with magnetic moments.

MOMENTUM The quantity of motion of a moving body.

QUANTIZED ENERGY LEVELS In an atom, they restrict the energy of electronic orbitals around the nucleus to assume only certain discrete magnitudes.

SPIN The intrinsic angular momentum of an atomic or subatomic particle.

Definitions adapted from
Oxford American English Dictionary
Encyclopaedia Britannica

Relevant reading

Anderson, M. H., Ensher, J. R., Matthews, M. R., Wieman, C. R., Cornell, E. A. (1995) Observation of Bose–Einstein condensation in a dilute atomic vapor. *Science* **269**: 198–201

Chu, S., Cohen-Tannoudji, C., Phillips, W.D. (1997) The Nobel Lectures. http://www.nobelprize.org/nobel_prizes/physics/laureates/1997

Davis, K. B., Mewes, M.-O., Andrews, M. R., *et al.* (1995) Bose–Einstein condensation in a gas of sodium atoms. *Physical Review Letters* **75**: 3969

Gavroglu, K., Goudaroulis, Y. (1989) *Methodological Aspects of the Development of Low Temperature Physics 1881–1956*. Concepts out of Contexts. Science and Philosophy Series. Kluwer Academic Publishers, Dordrecht

Mendelssohn, K. (1966) *The Quest for Absolute Zero*. Weidenfeld & Nicolson, London

Roller, D. (1950) *The Early Development of the Concepts of Temperature and Heat. The Rise and Decline of the Caloric Theory*. Harvard University Press, Cambridge

Photo credits

p 3   Marlene Dietrich in the film *Desire*, courtesy of the Everett Collection

p 5   The inside of the ALICE TPC © 2004 CERN, Author: Michael Hoch

p 5   The Sun's chromosphere, courtesy of JAXA/NASA

p 5   View of the Crab Nebula © 2010 NASA, ESA, ASU, J. Hester

p 5   Caravan in the Sahara Desert © Galyna Andrushko – Fotolia

p 5   Baby © Oleg Kozlov – Fotolia

p 5   Penguins © Bernard Breton – Fotolia

p 5   Photograph from *Liquid Helium*, *Superfluid* by Alfred Leitner

p 5   Liquid nitrogen © Zirafek – Fotolia

p 5   Visualization of the Cosmic Microwave Background, corresponding to a temperature of ~2.72K © NASA/WMAP Science Team

p 5   Data confirming the discovery of a Bose–Einstein condensate of a gas of rubidium atoms © 1995 NIST, JILA, CU-Boulder

p 5   Artist view of laser-cooling setup © 2010 ICFO-Digivision

GENTLE

Optical tweezers

GENTLE

Go – from the Japanese word igo, meaning "encircling game" – is a strategy board game that originated in China approximately 3500 years ago. The game starts with an empty grid set on a wooden board. Players then take turns to place black and white stones on the intersections of the crossing lines. The objective of each player is to capture the stones of the opponent. A stone – or chain of connected stones – is captured if all its four degrees of freedom (the four surrounding intersections) are occupied by an opponent's stones. Despite these simple rules, go is a game of extraordinary complexity and beauty. A player in any typical game among experts has an average of a few hundred choices per move, making strategies strongly tied to intuition, experience, and pattern recognition.[1] The players' skills strengthen by learning to identify a certain balance between the territory they give away and the force they exert on the opponent. It is a gentle-strategy game.

The lack of skillfulness typically results in pushing the opponent's beads instead of encircling them. Successful encircling calls for a delicate, wise balance between force and territory, limiting and yielding, pushing and letting go. A suitable way to go: attack from the corners. Continue along the sides. Move into the center. Trap. The goal is to constrain the freedom of the opponent applying the minimum amount of force.

In Nature, a similar balance sustains the glide of a seagull on a current of air. The form of the wings separates the flow of air molecules so that moving air molecules passing underneath exert a lifting force larger than the downward pressure of those passing above. The resulting force is just enough to compensate gravity and keep the bird in a comfortable flight.

In spirit, a similar balance is the principle behind optical trapping. Photons can exert tiny – but observable – forces by exchanging momentum with very small particles. Without a good strategy, though, photons can only push. If they are to "capture" a particle, photons need to "attack from the corners, continue along the sides, and then move into the center." The photon flow has to be shaped to create a balance of opposite forces on the particle. However, particles do not have wings, and therefore the flow needs to be shaped – focused – with the help of a lens.

This strategy allows the study of the composition, structure, and function of delicate microscopic objects like living cells, organelles, or DNA strands, exerting the right force to trap them under the microscope and leaving them unharmed.

---

[1] On a 19 x19 matrix, there are at least 10^{480} possible games of go. Shannon gave a number of 10^{120} possible games for chess. The visible Universe is estimated to have between 10^{79} and 10^{81} atoms.

Unity
is strength.

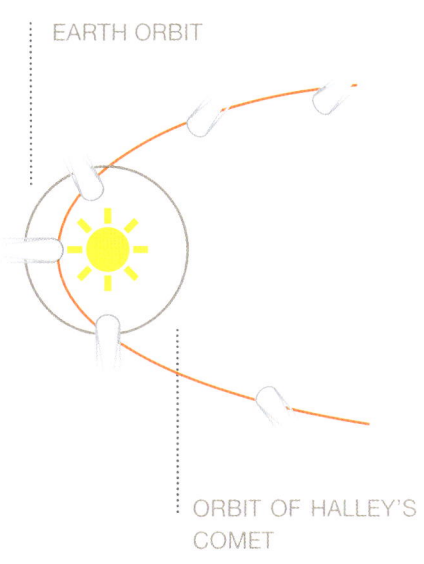

EARTH ORBIT

ORBIT OF HALLEY'S
COMET

Solar photons push the
molecules in the gases
of a comet's tail, and
make them point away
from the Sun.

But the force exerted
by each photon

is minuscule.

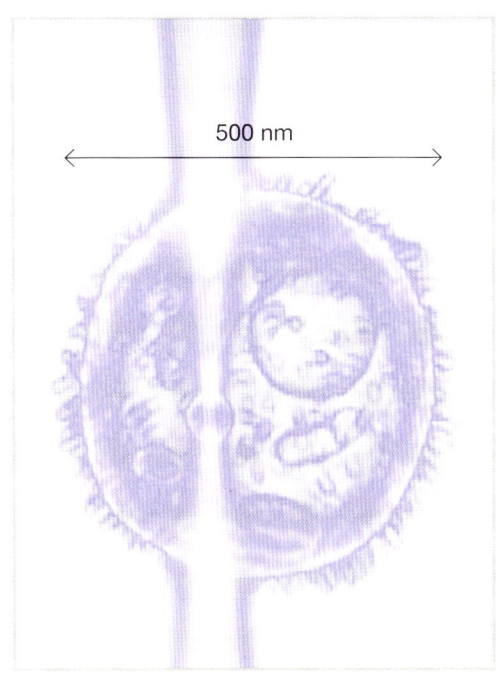

500 nm

Can light exert controllable
gentle forces on very small,
possibly living, objects?

Can it hold them, stretch
them, and move them, without
harming them?

The force of radiation pressure from a **laser** was illustrated in a beautiful experiment conducted in 1970. With an argon ion laser of one watt power and **wavelength** of half a micron focused upon a dielectric sphere of the same radius, a radiation pressure force of 6.6×10^{-10} newtons was measured. Macroscopically, the radiation pressure can be understood as the pressure exerted by the photons hitting and reflecting from the sphere. Microscopically, the momentum exchange between photons and matter takes place through absorption and re-emission of photons through transitions between the electronic levels of molecules and atoms. Such a geometry allows small particles to be trapped and guided.

A key landmark in the realization of optical tweezers was achieved in 1986. In standard geometries, the angle formed by the incident photons and the surface of the sphere is determined by the **numerical aperture** of the lens that focuses the beam. The relative **refractive** index between the sphere and the surrounding medium determines how many photons reflect on the sphere – hitting it in one direction – and how many photons are refracted through it.

This relation between scattering and **refraction** causes a sphere of higher refractive index than its surroundings to be pulled into the strongest part of the beam and kept trapped. This effect is reliant therefore on a strong optical field gradient.

Nowadays, for biological and medical applications, the laser power can be minimized to avoid compromising the integrity of the sample. At the nanoscale, nanoparticles, nanostructures, and nanoapertures can be used to concentrate light in really small regions and trap directly nano-sized objects, like virus or single molecules by creating highly localized optical field gradients that go beyond that from standard optical components. Applications are the study of infection at the single-virion level or manipulation and assembly of a sample at the single-molecule level.

So good-natured

as to deal with small delicate living objects, leaving them unharmed, almost untouched.

With the right balance between a small object and its medium, light can behave as a minute tweezer, with no more force than necessary, with outstanding accuracy and gentleness beyond the reach of any material mechanical system.

In the picture, a bacterium measuring 0.000 0005 meters floats in a stream of photons.

Scientific and technical advisors

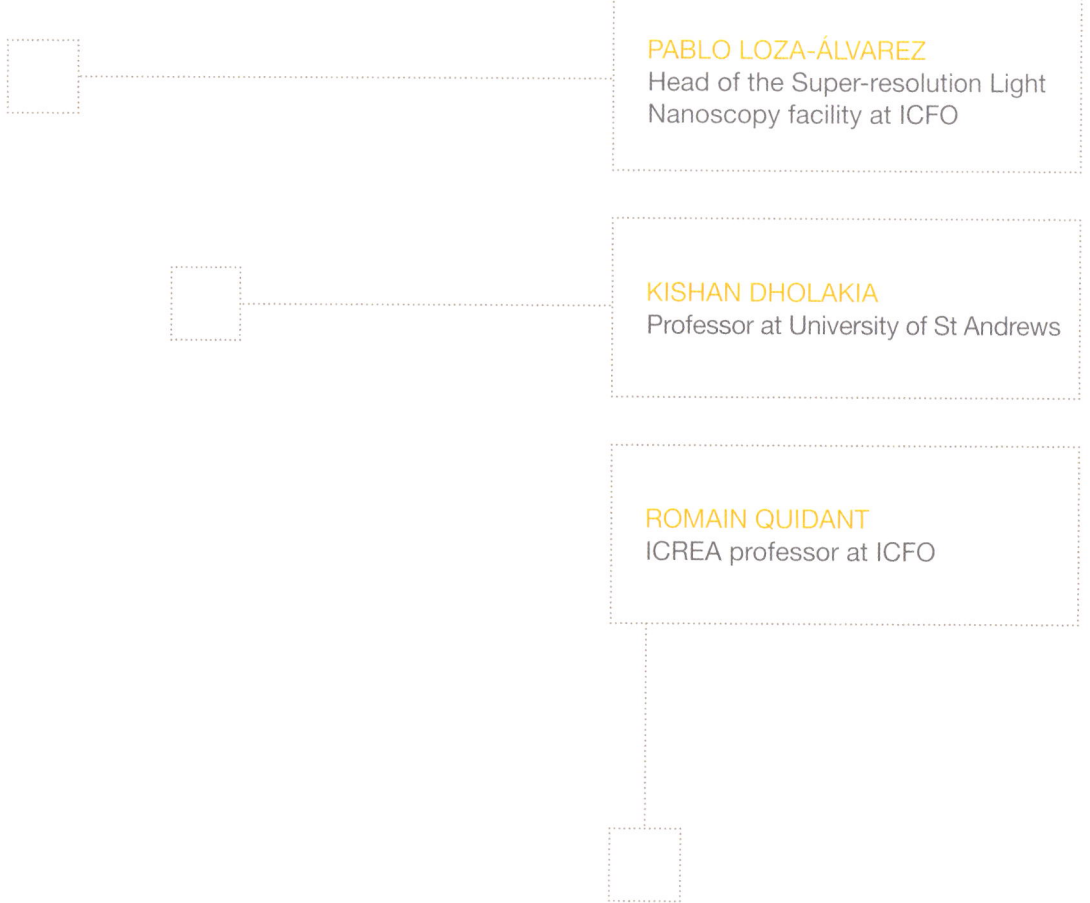

PABLO LOZA-ÁLVAREZ
Head of the Super-resolution Light
Nanoscopy facility at ICFO

KISHAN DHOLAKIA
Professor at University of St Andrews

ROMAIN QUIDANT
ICREA professor at ICFO

Glossary

LASER Acronym for Light Amplification by Stimulated Emission of Radiation. A laser produces a highly coherent, highly directional, and nearly monochromatic beam of light, by stimulating atoms or molecules to emit light at particular wavelengths and amplifies that light, typically producing a very narrow beam of radiation.

NUMERICAL APERTURE Dimensionless number that characterizes the range of angles over which a system can accept or emit light. It is the sine of the vertex angle of the largest cone of meridional rays that can enter or leave the lens, multiplied by the refractive index of the medium in which the vertex of the cone is located. The numerical aperture is the main parameter that determines the resolving power of a lens.

REFRACTION The bending of oblique incident rays as they pass from a medium having one refractive index into a medium with a different refractive index.

REFRACTIVE INDEX Measure of the bending of a ray of light when passing from one medium into another.

WAVELENGTH Electromagnetic energy is transmitted in the form of a sinusoidal wave. The wavelength is the physical distance covered by one cycle of this wave.

Definitions adapted from
Oxford American English Dictionary
Encyclopaedia Britannica

Relevant reading

Ashkin, A. (1970) Acceleration and trapping of particles by radiation pressure. *Physical Review Letters* **24**: 156

Ashkin, A., Dziedzic, J. M., Bjorkholm, J. E., Chu, S. (1986) Observation of a single-beam gradient force optical trap for dielectric particles. *Optics Letters* **11**: 288–290

Ashkin, A. (2011) How it all began. *Nature Photonics* **5**: 316–317

Juan, M. L., Righini, M., Quidant, R. (2011) Plasmon nano-optical tweezers. *Nature Photonics* **5**: 349–356

Grier, D. G. (2003) A revolution in optical manipulation. *Nature* **424**: 810–816

Clarke, A.C. (1972) *The Wind from the Sun*. Harcourt Brace Jovanovich, San Diego

Photo credits

p 11   Artist conception of a cell and bacteria trapped in optical tweezers © ICFO-Digivision

p 12   Playing go in Japan © 2010 A. Davey

SHARP

Laser surgery

SHARP

SHARP

Incisive, brilliant, discriminating, precise? Elementary: Sherlock Holmes. The master of observation and deduction, unmistakable in his cape, pipe, and magnifying glass, yet able to penetrate any layer of society when camouflaged in one of his thousands of disguises. Certainly not the tidiest man in his personal habits, but the most meticulous when dissecting a case, with an innate ability to single out guilt from innocence, discern minute details, avoid external bias, and operate at full power without drawing attention. A true detective of surgical methods. Not entirely fictional, for Sir Arthur Conan Doyle's inspiration for this character was his professor at the Edinburgh Medical School, the Scottish surgeon Dr. Joseph Bell – apparently, an extraordinary lecturer and physician, whose sharp eye and inference powers were frequently requested by the police in the toughest investigations.

During Conan Doyle's lifetime, especially during the last decades of the nineteenth century – while he still combined writing with medical practice – surgery became a well-established medical discipline. Two fundamental advances were involved: anesthesia and sterilization. In the 1840s, the discovery of ether and chloroform allowed doctors to overcome the barrier of pain, and surgeons were able to venture deeper, performing more elaborate operations. Long story short, in the 1860s, John Lister implemented sterilization for fighting infection, following the steps of the unfortunate

Ignaz Semmelweis, and the groundbreaking experiments of Louis Pasteur, which led to the confirmation that infection was caused by microorganisms.

Modern surgery has continued to advance in the search for improved methods to fight pain and infection, while striving to minimize invasiveness, recovery time, and secondary effects. Instruments grew sharper and gentler in order to deal with smaller and more delicate objects. The advance of microscopy and computer-assisted manipulation now helps surgeons to go beyond the power of their eyes and hands to dissect and solve the most complex cases, sometimes hidden behind layers of tissue, as in the case of a blocked coronary artery or a leak of cerebrospinal fluid. But, what kind of tool do they use to be able to cross, unnoticed, through several layers of tissue? And once there, what device gives them the sharp precision necessary to hit a microscopic target and leave the surroundings unharmed? Elementary: laser light.

Laser surgery is a common practice in oncology (e.g. tumor ablation), dissection (e.g. cutting and coagulation of small blood vessels), ophthalmology (e.g. myopia correction, retinal detachment), dermatology (e.g. skin tumors), endoscopic surgery, or otorhinolaryngology, or cosmetics. Getting deeper, to modify without damaging the pattern of subcellular structures and neural connections is certainly a job for the sharpest detective with innumerable disguises.

Cutting through
an object without
damaging the
surrounding area is
not an easy task.

Collateral damage
depends on the relative
size of and roughness
of the knife.

The challenge is to find incisive sources of energy
to reduce collateral damage at the smallest scale.

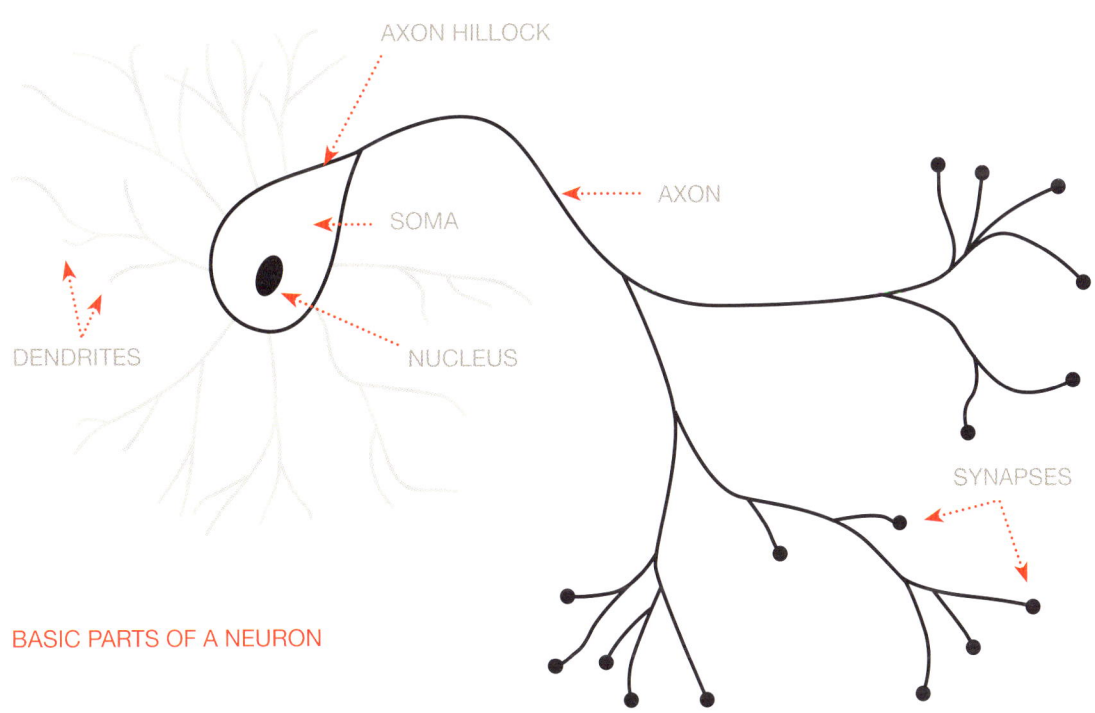

AXON HILLOCK

AXON

SOMA

DENDRITES

NUCLEUS

SYNAPSES

BASIC PARTS OF A NEURON

SHARP

Laser surgery

There are many different mechanisms by which **laser** light can dissect tissue. The dominant mechanism that determines such interaction depends on the type of tissue, the **wavelength** of the light, the delivered intensity, the duration and repetition of the illumination, and the beam shape. Lasers perform as great tools because all these parameters can be carefully controlled.

In conventional laser surgery, the dominant interaction is photothermal: the energy of the photons is absorbed by **chromophores**, and then the molecules relax by converting this energy into heat localized nearby the laser spot. This results in several effects in the tissue, ranging from vaporization to coagulation. As a consequence, and since the laser spot size can be small – a few millimeters or less – lasers can produce very precise cuts. In other circumstances, because of the several processes involved, they can also be used to weld and cauterize wounds.

Apart from photothermal effects, other common interaction mechanisms that are involved in laser cutting include photochemical reactions, photoablation, photodisruption, and **plasma-induced ablation** (PIA). In all of them, the type and extent of the produced incision is different.

In particular, using the process of PIA, it is possible to perform nanosurgery in living cells and tissues without altering the surrounding target volume. This PIA can occur by tightly focusing ultrashort (femtosecond) laser pulses and taking advantage of a nonlinear absorption process, in which the incident optical intensity is sufficiently high to stimulate the simultaneous absorption of multiple photons. These photons generate a cascade ionization process that results in a highly localized ablation, which can be smaller in size than the focal volume. As thermal effects are not present, collateral damage is strongly minimized. Furthermore, as normally **infrared** wavelengths are used, the target can be placed up to a few hundred microns inside the sample.

Like knives, **laser beams can be sharpened.**

Unlike regular knives, they let you **trim their millimetric size a thousand times,** then concentrate their power another thousand times to perform the sharpest dissections on the intracellular scale.

Ultrashort pulses (150 fs) in the infrared performing a cut on a *C. elegans* head neuron. The neuron thickness is approx. 300 nm.

Scientific and technical advisors

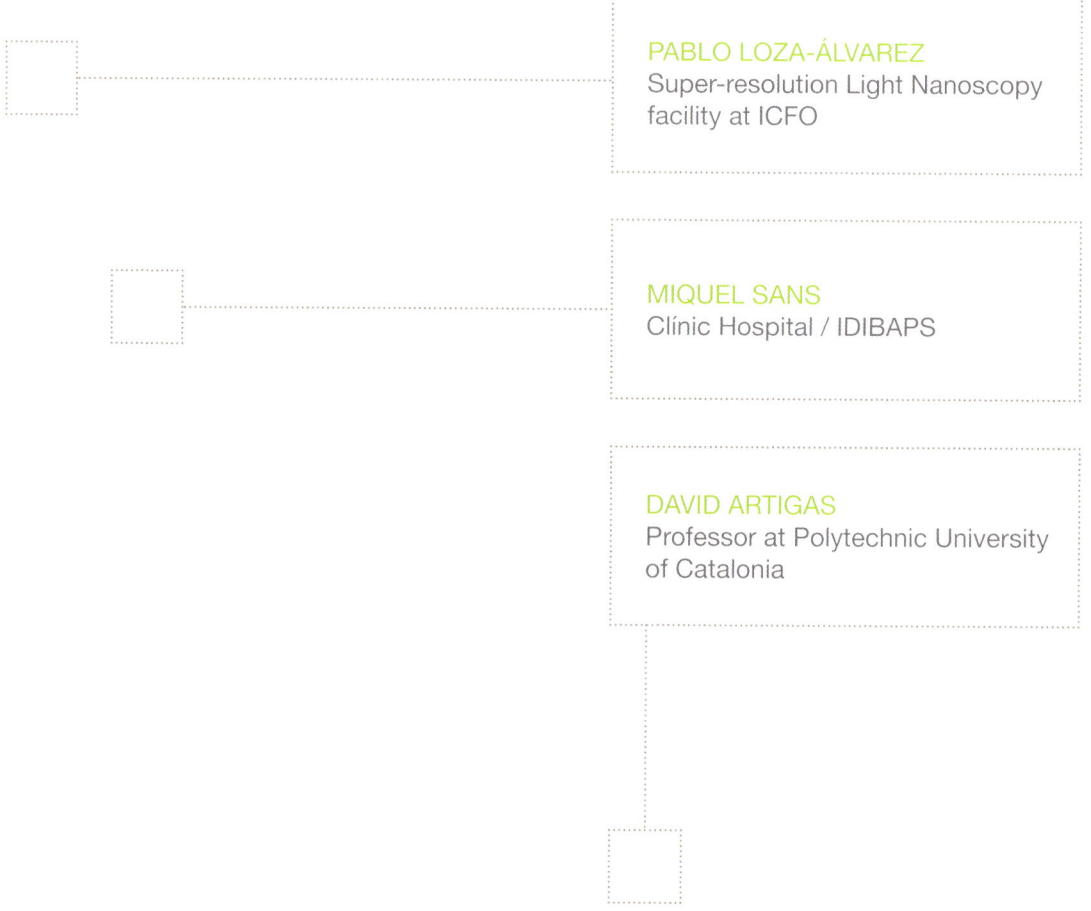

PABLO LOZA-ÁLVAREZ
Super-resolution Light Nanoscopy facility at ICFO

MIQUEL SANS
Clínic Hospital / IDIBAPS

DAVID ARTIGAS
Professor at Polytechnic University of Catalonia

Glossary

CHROMOPHORE Part of an organic molecule that causes it to be colored. Electrons at the molecular orbitals can absorb energy from light over a certain range of wavelengths in the visible region; transmission or reflection of the remainder of the light gives rise to the observed color of the compound.

FREQUENCY The number of waves that pass a fixed point in a unit of time. Usually frequency is expressed in the hertz unit, equal to a unit of one cycle per second.

INFRARED The portion of the electromagnetic spectrum extending from the red end of the visible region to the microwave range. The infrared range is usually divided into three regions: near-infrared, from 700 nm to 2 μm; medium-infrared, from 2 μm to 4 μm; and far-infrared, from 4 μm to 1 mm.

LASER Acronym for Light Amplification by Stimulated Emission of Radiation. A laser produces a highly coherent, highly directional, and nearly monochromatic beam of light, by stimulating atoms or molecules to emit light at particular wavelengths and amplifies that light, typically producing a very narrow beam of radiation.

PLASMA-INDUCED (ABLATION) The electric field of an intense laser can ionize the molecules of a sample to form a plasma. Such a mixture of free electrons and ionized molecules can absorb incoming radiation of any wavelength, since the energy levels of plasma are not generally quantized.

WAVELENGTH Electromagnetic energy is transmitted in the form of a sinusoidal wave. The wavelength is the physical distance covered by one cycle of this wave; it is inversely proportional to frequency.

Definitions adapted from
Oxford American English Dictionary
Encyclopaedia Britannica

Relevant reading

Chung, S. H., Mazu, E. (2009) Surgical applications of femtosecond lasers. *Journal of Biophotonics* **2(10)**: 557–572

Goetz, T. (2014) *The Remedy: Robert Koch, Arthur Conan Doyle, and the Quest to Cure Tuberculosis.* Gotham Books, New York

Ellis, H. (2009) *The Cambridge Illustrated History of Surgery.* Cambridge University Press, Cambridge, New York

Niemz, M. H. (2004) *Laser–Tissue Interactions.* Springer, Leipzig

Photo credits

p 20   Basil Rathbone as Sherlock Holmes in the *Hound of the Baskervilles*, courtesy of the Everett Collection

p 21   Fluorescence image of the head neurons of a *C. elegans* © ICFO

FOCUS

Super-resolution microscopy

FOCUS

Focus is the principle behind vision. First, attention is focused on an object. Then a lens in the eye collects and focuses onto the retina (the screen of the eye) the light reflected or emitted by every point of the selected object. Finally, the information thus gathered is processed by the brain to construct a meaningful image.

In order to offer a distinct, significant, and almost instantaneous image of the world around us, the brain is obliged to focus on a narrow subset of relevant information. The input provided by our senses is selectively picked out, since processing the whole set of data would exceed the brain's computational capacity. (That might explain why evolution did not burden us with the power of microscopic vision: it would mean way too much information to process!)

To see small details beyond the limits reached by evolution, the aid of more-powerful tools is needed. Larger and stronger lenses in microscopes overcome the capacity of the human eye's lenses to collect and separately focus on the screen the light that arrives from two very near points of the sample. As a result, features of a structure indiscernible to the naked eye appear clearly differentiated. However, focus (and hence vision) is limited by a fundamental fact: light is a wave.

The wavelength of light involved in vision is approximately 500 nm. Points spaced apart by 200–300 nm cannot be distinguished by any conventional lens, for diffraction will cause two overlapping blurred spots instead of two well-differentiated image points. This fundamental limit for resolution is known as the diffraction limit of light. Scientists are currently busy trying to beat it using a variety of tricks, including the use of certain exceptional materials – metamaterials – to fabricate the future so-called super-lenses. However, current technology can already achieve an extraordinary control of the interaction of light and matter at the nanoscale, providing a set of alternative methods to image details under the diffraction limit of light. These techniques constitute the field of super-resolution microscopy.

To avoid the overlapping of light arriving from two close emitters, one group of super-resolution techniques is based on the selective activation or deactivation of light coming from different points of the sample. The positions of two close points can thus be discerned by controlling which of them is allowed to glow at a given time, or in a particular color. It is like listening to overlapping conversations at a cocktail party, just by having the power to block or pause some of them at will, or having different conversations held in different languages. A second group of techniques concentrates light on small (nanofabricated) tips to scan the sample point by point. Images are generated by bringing the tip very close (about 10 nm) to the sample, and then transducing the detected signal into visual information.

With an active effort to focus on relevant details (and a great deal of computational resources) super-resolution tools can provide images of bio-structures, single molecules, and other details of the nanoscopic world worth focusing on, with a striking resolution of 10–30 nm.

A lens in the eye focuses at the retina the light coming from every point of an object.

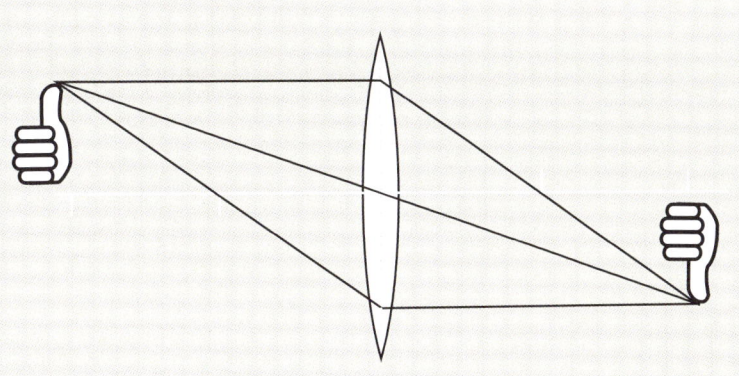

Lenses in magnifiers and microscopes are able to focus on more details, and offer a point-to-point image of very small areas.

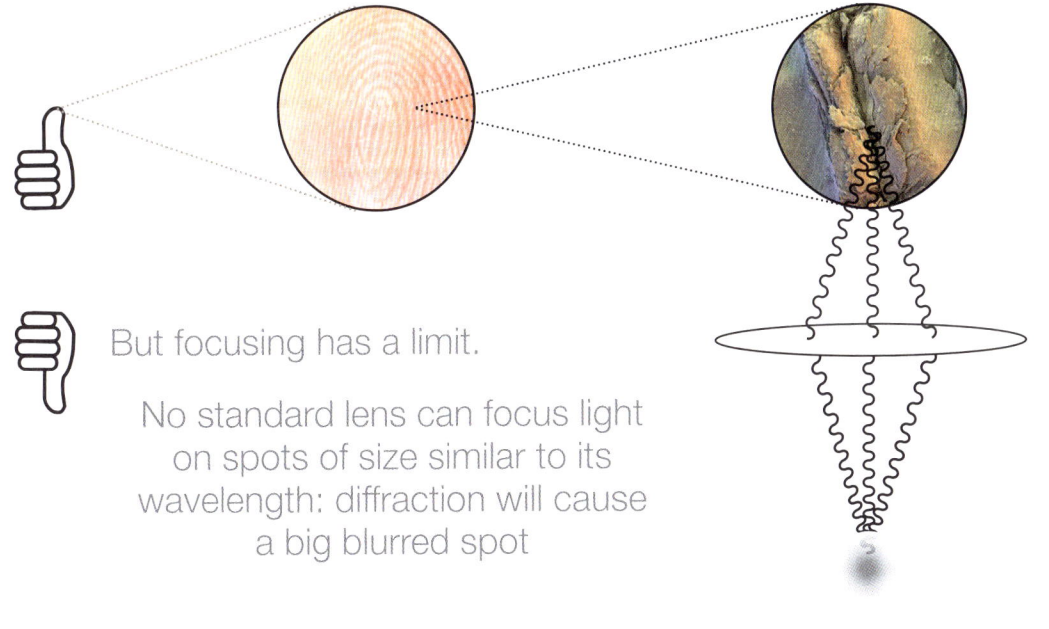

But focusing has a limit.

No standard lens can focus light on spots of size similar to its wavelength: diffraction will cause a big blurred spot

The challenge is where to focus to perceive at a scale in which we cannot see.

FOCUS

Super-resolution microscopy

Diffraction imposes a fundamental limit in microscopy: available lenses cannot resolve points spaced by less than $d = \lambda/(2NA)$, where λ is the **wavelength** of light and NA is the **numerical aperture** of the lens. For visible light, the diffraction limit is approximately 200–300 nm.

Super-resolution microscopy is a set of techniques for the generation of images in which points spaced by less than the resolution limit are distinguishable. Different strategies are being pursued to achieve super-resolution.

A variety of methods based on **fluorescence** are used to image subcellular details, e.g. cell organelles, protein structures. These methods rely on the detection of light emitted by a sample in which some molecules have been treated with **fluorophores**. **Photoactivation** and **photoswitching** allow an active control of the light emitted by targeted molecules.

Such control avoids the overlap of diffraction spots corresponding to two close emitters. The final image is a reconstruction of the most likely position of each identified emitter. An example would be STORM – Stochastic Optical Reconstruction Microscopy. Other fluorescence techniques reduce the diffraction spots through a special design of the optical response of the sample, selectively deactivating fluorophores. For example, in STED – Stimulated Emission Depletion-microscopy, the fluorophores are switched off sequentially by simultaneously scanning a focused excitation laser and a donut shaped depletion laser. Resolutions of 20–30 nm can be readily achieved, in some cases reaching 1 nm.

An alternative approach to circumvent the diffraction limit is the use of ultrasharp tips that provide enhanced light emission in a very reduced volume. Etched optical fibers or nano-antennas probe the optical response of a sample at the single-molecule level through a point-by-point scanning of the surface. In addition to providing information about the structure, the rendered image may convey biochemical insight on composition too. These techniques attain resolutions of about 15 nm and are well suited to study biological samples and processes, such as those occurring at the cell membrane.

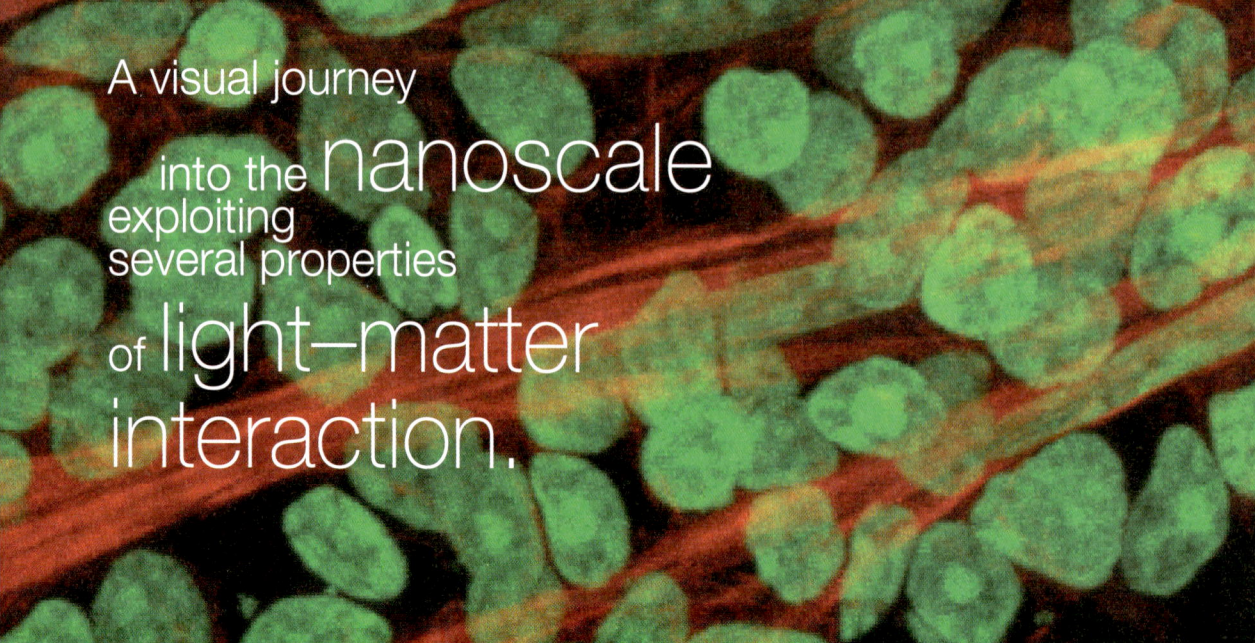

A visual journey into the nanoscale exploiting several properties of light–matter interaction.

Reflection and transmission of light make possible sight and standard microscopes.

Emission of light from targets helps in visualizing details...

...even **beyond the diffraction limit of light...**

using reconstruction techniques,

or interacting directly with **optical resonances** of nanoscopic objects.

1 mm

1 μm

100 nm

10 nm

1 nm

Tropical frog. Naked eye.

10μm

Head of a *C. elegans*. Standard microscopy.

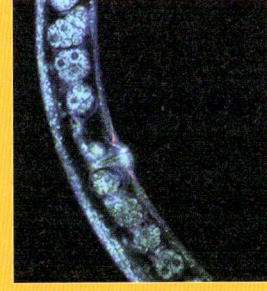
Embryos of a *C. elegans*. Fluorescence confocal microscopy.

Body wall muscles of a *C. elegans*. Second-order harmonic generation.

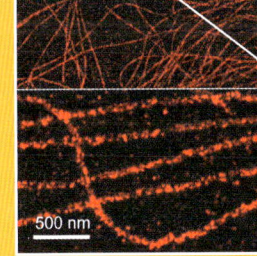
500 nm

Cell microtubules. Stochastic Optical Reconstruction Microscopy – STORM.

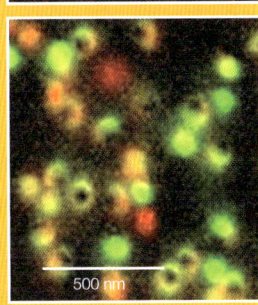
500 nm

Single fluorescent molecules imaged by a nano-antenna probe.

Scientific and technical advisors

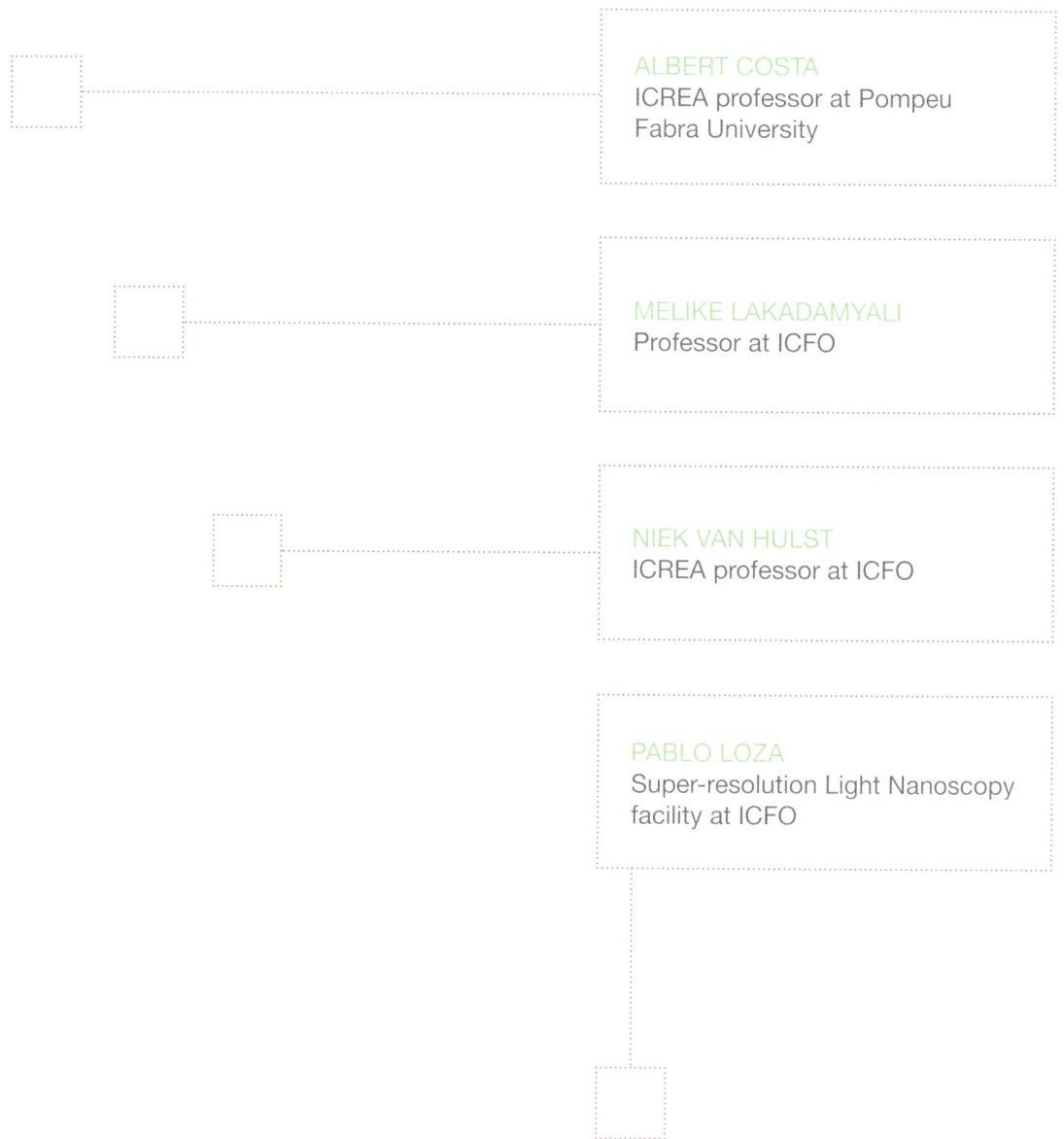

ALBERT COSTA
ICREA professor at Pompeu
Fabra University

MELIKE LAKADAMYALI
Professor at ICFO

NIEK VAN HULST
ICREA professor at ICFO

PABLO LOZA
Super-resolution Light Nanoscopy
facility at ICFO

Glossary

....

DIFFRACTION As a wavefront of light passes by an opaque edge or through an opening, secondary weaker wavefronts are generated, apparently originating at that edge. These secondary wavefronts will interfere with the primary wavefront as well as with each other to form various diffraction patterns.

FLUORESCENCE Visible or invisible radiation emitted by certain substances as a result of incident radiation of a shorter wavelength.

FLUOROPHORE A fluorescent chemical compound that, upon light excitation, can re-emit light of a different wavelength.

NUMERICAL APERTURE Dimensionless number that characterizes the range of angles over which a system can accept or emit light. It is the sine of the vertex angle of the largest cone of meridional rays that can enter or leave the lens, multiplied by the refractive index of the medium in which the vertex of the cone is located. The numerical aperture is the main parameter that determines the resolving power of a lens.

PHOTOACTIVATION The process of activating a substance by means of radiant energy and especially light.

PHOTOSWITCHING On/off switching of fluorophores by light.

WAVELENGTH Electromagnetic energy is transmitted in the form of a sinusoidal wave. The wavelength is the physical distance covered by one cycle of this wave; it is inversely proportional to frequency.

....

Definitions adapted from
Oxford American English Dictionary
Encyclopaedia Britannica

Relevant reading

Frankel, F. (2009) *No Small Matter.* Harvard University Press, Cambridge, MA

Garcia-Parajo, M. F. (2008) Optical antennas focus in on biology. *Nature Photonics* **2**: 201–203

Hell, S. W. (2007) Far-field optical nanoscopy. *Science* **316**: 1153–1158

Won, R. (2009) Eyes on super-resolution. *Nature Photonics* **3**: 368–369

Croft, W. J. (2006) *Under the Microscope: A Brief History of Microscopy.* World Scientific, Singapore

Photo credits

p 27 Digital imprint © thawats – Fotolia
p 27 Electron microscope image of a human fingerprint ridge © Hutch Media/Justin Thomas
p 28 Confocal maximal projection of a 16 µm section of mouse intestine © ICFO, author: Jordi Andilla
p 29 Tropical frog © Christophe Fouquin – Fotolia

VIRUS ATTACK

But don't panic!

VIRUSATTACK

In classical science fiction, alien attacks are perpetrated by oddly shaped creatures from outer space. Since the turn of the century, the genre has taken a more disturbing turn: invaders are earthly viruses. Although fictional, these new plots are based on a threat already experienced by a reliable majority: influenza, a well-known virus as small as 100 nm, infects millions of people every year, spreading the flu worldwide via airborne or direct transmission.

One of these fictional plots could feature Sarah, a young virologist facing the spread of an unknown virus threatening humanity. After a first observation of the symptoms and the evolution of patients, she sits at the microscope to focus on virus–cell interactions. Such interactions are fundamental to initiate an infection, since viruses need the machinery of the cell to be replicated and expressed.

Hers is a first-class standard microscope. It has the best components and offers images of the best possible resolution, but it is limited by the basic laws of light ray transmission. These are the same principles that explain rainbows and lenses, and were already beautifully summarized by Pierre Fermat in The Principle of Least Time: "Out of all the possible paths that light might take to go from one point to another, it takes the path that requires the shortest time." Since Fermat, scientists have learnt that this principle is not completely general, but they have put forward modern refined versions that are.

When Sarah switches on the white lamp under the sample dish, the rays take Fermat's path, bending and refracting as they travel through different mediums in a labyrinth of crystal lenses, air, and mirrors. As a result, a greyish image of deformed cells reaches her eyes, indicating the infection is on course. She needs to understand how the virus is attacking; otherwise she will not be able to find a way to block or divert its maneuvers before every human dies. However, the image of the cells, an amalgam of indistinct moving grey shapes, does not give enough contrast to clearly identify the agents involved in the invasion. What is the entry pathway of the virus once it lands on the cell's membrane? Which cell structures transport the virus into the nucleus? In the case of influenza, she knows that the pathway is a protein called clathrin and that highway-like structures called microtubules are responsible for the transport. But it is hopeless for her to point out the infection mechanisms of the new virus, unless she can use new tools to get a better image.

The film reaches the point of maximal suspense. Fermat did not know about fluorescence – or any quantum behavior of light – but the film scriptwriters did, and gave the story a final quantum leap. Sarah sends the sample to the next-door lab, Fluorescence Microscopy, led by Elsa and Jan – Sarah's sister and her boyfriend – only to discover they are both infected by the virus. Fluorescence microscopy uses bright colors to label virons and cell structures, which can be clearly visualized when illuminated with laser light of the right color. Elsa has just enough strength to dye the samples. Jan turns on the lasers. As a result, brilliant colorful structures reveal the activity of the different cell parts linked to the entry of the virus.

In this story, a protein in the cell membrane would be identified as the entry path. Sarah would produce a drug to inhibit the action of the protein in time to save Elsa, Jan, and everybody else.

In the real world, things might take a bit longer.

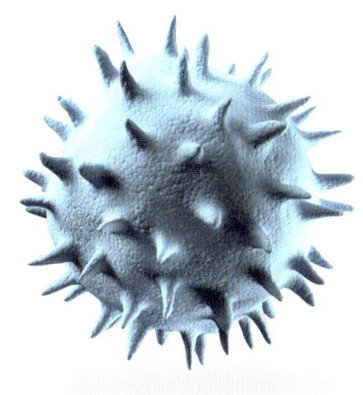

Cell mechanisms can be fooled by external enemies…

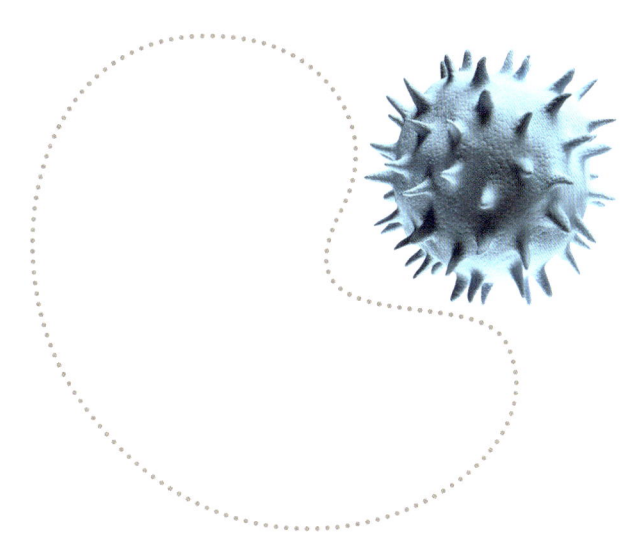

…and perform as **perfect allies for an alien invasion.**

How does this happen?

The relevant movements **occur** inside a region of a billionth of a meter!

VIRUS **ATTACK!**

But **don't** panic

The main image on the right shows an artistic impression of an influenza virus entering a cell. The images at the bottom correspond to different time points from a **fluorescence** microscope imaging of a viral entry. The image allows us to visualize that when the virus (red) binds to its receptors on the cell membrane, specific proteins such as clathrin (green) are recruited to the binding site to deliver the virus into the cellular interior. The virus is then trafficked inside **endosomes** along cytoskeletal tracks towards the cell nucleus. As the virus approaches the nucleus, the endosomes also undergo a transformation that includes changes to their protein content and pH, and finally once the endosomal pH becomes acidic enough, the viral membrane fuses with the endosomal membrane to allow the release of the viral genome.

To obtain such detailed information on viral entry, each single virus must be visualized by incorporating **fluorophores** into the viral membrane that make the virus glow a bright red color when illuminated with laser light of a certain **wavelength** (633 nm). The virus can be followed in real time, from the time it binds to the cell membrane till the time it releases its genome into the cell, by using time-lapse imaging and single-particle tracking. Cellular components, like clathrin or endosomes, can be labeled with fluorescent proteins that glow a bright blue or yellow color when illuminated with **laser** light of different wavelengths (457 nm or 532 nm). The fluorescence emission from the virus and the cellular components can be split and monitored simultaneously in separate channels in a multi-color imaging session to determine their interactions.

Viruses invade organisms to replicate, but the exact entry mechanisms used by many of them remain unclear.

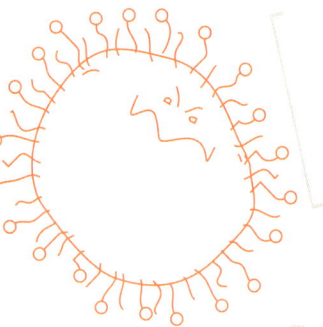

In less than an hour, a single virus makes it from a cell's membrane to its nucleus. Some days later the viral RNA has been replicated by the cell's machinery several thousand times.

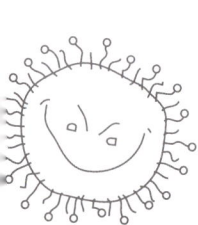

Fluorescence imaging aids in the development of new treatment by revealing the details of infection mechanisms used by viruses.

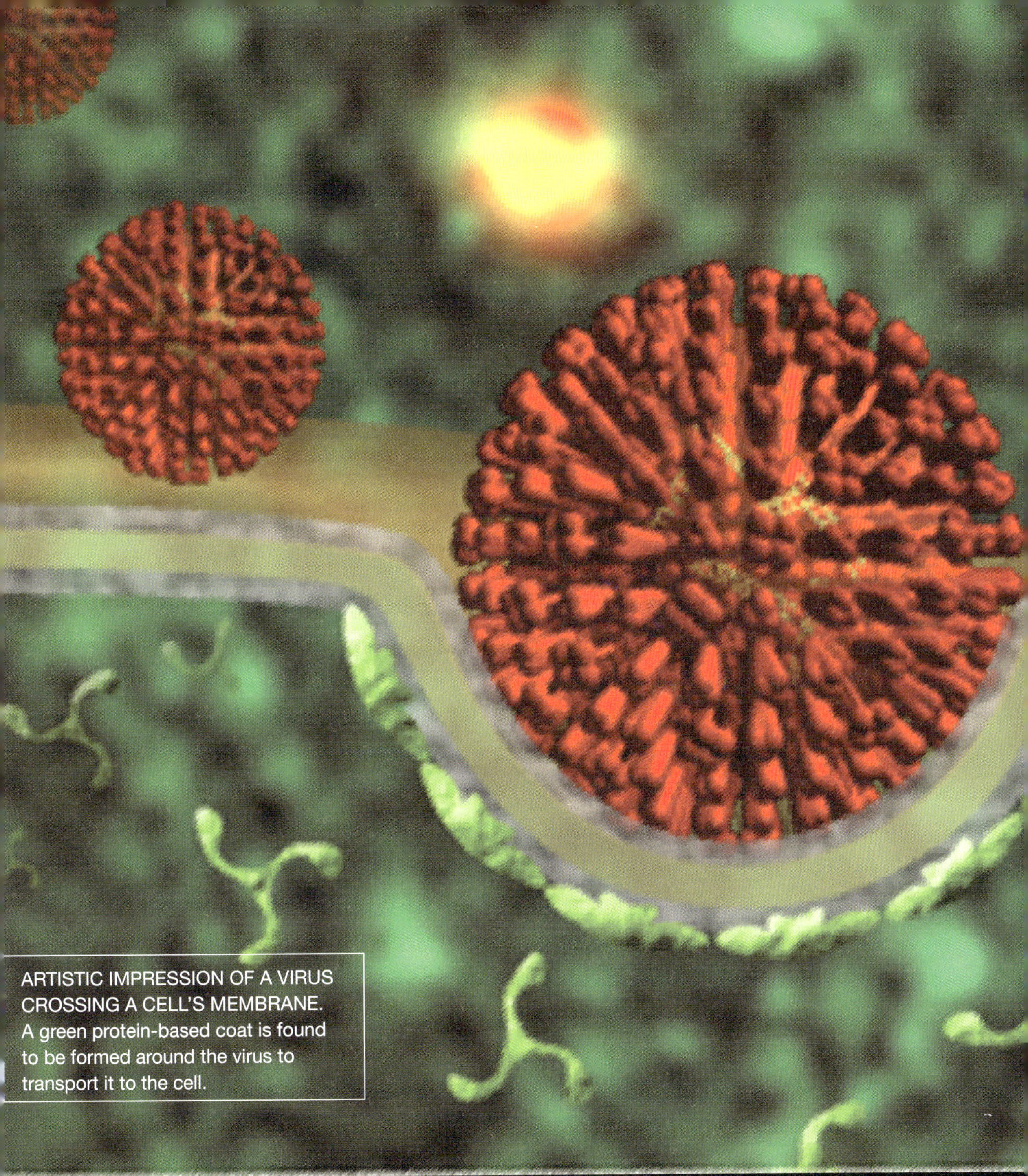

ARTISTIC IMPRESSION OF A VIRUS CROSSING A CELL'S MEMBRANE. A green protein-based coat is found to be formed around the virus to transport it to the cell.

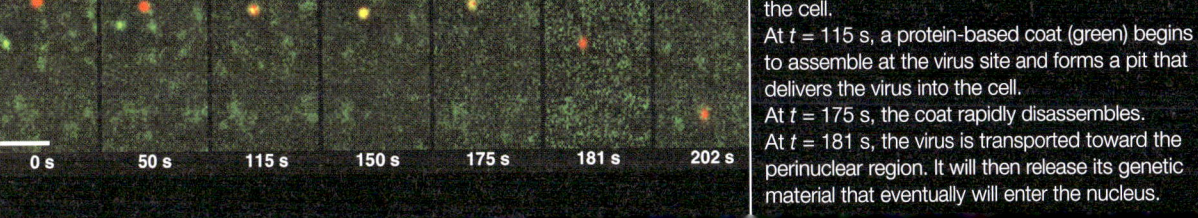

Scale bar, 10 µm. At t = 0 s, the virus (red) binds to the cell.
At t = 115 s, a protein-based coat (green) begins to assemble at the virus site and forms a pit that delivers the virus into the cell.
At t = 175 s, the coat rapidly disassembles.
At t = 181 s, the virus is transported toward the perinuclear region. It will then release its genetic material that eventually will enter the nucleus.

0 s   50 s   115 s   150 s   175 s   181 s   202 s

Scientific and technical advisors

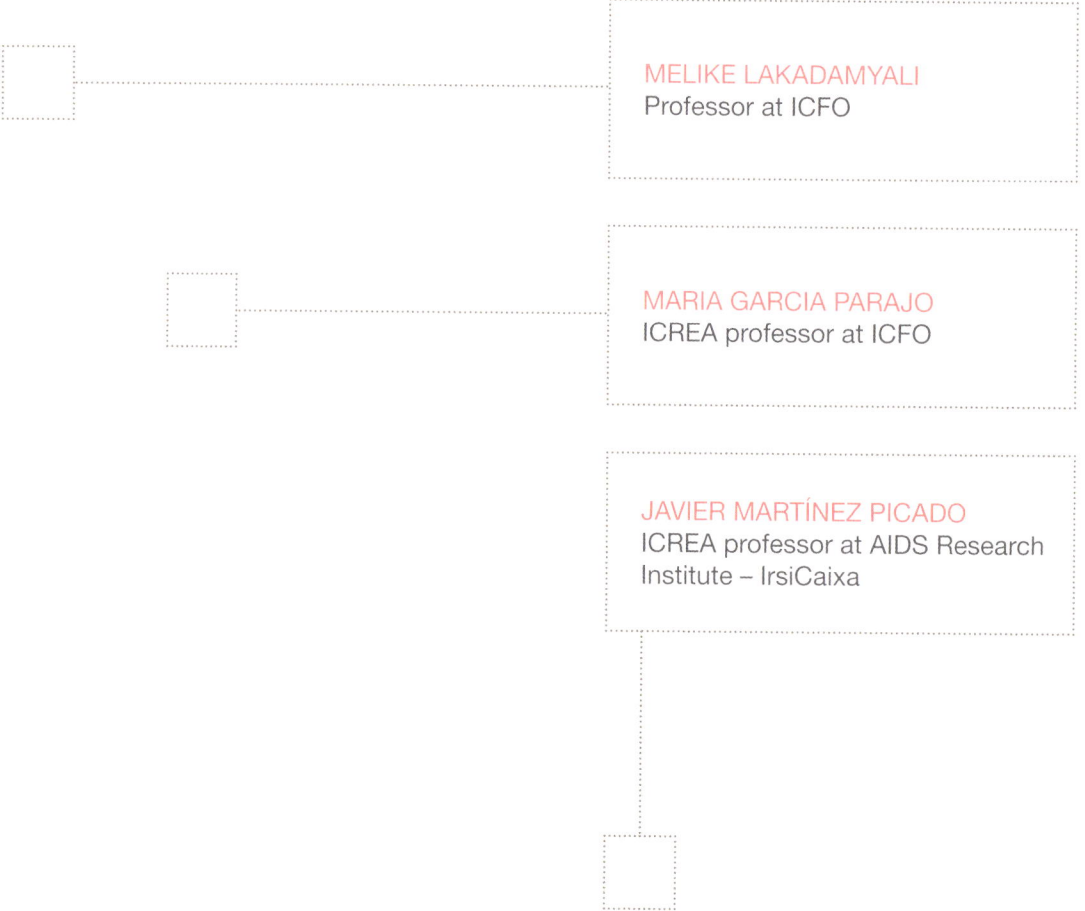

MELIKE LAKADAMYALI
Professor at ICFO

MARIA GARCIA PARAJO
ICREA professor at ICFO

JAVIER MARTÍNEZ PICADO
ICREA professor at AIDS Research
Institute – IrsiCaixa

Glossary

ENDOSOME Cell organelle involved in the reception and trafficking of new material to the cell interior.

FLUORESCENCE Emission of radiation by certain substances as a result of an incident radiation of a shorter wavelength, such as X-rays or ultraviolet light. The atoms are excited by the incident radiation and re-emit almost immediately.

FLUOROPHORE A fluorescent chemical compound that can re-emit light upon light excitation.

LASER Acronym for Light Amplification by Stimulated Emission of Radiation. A laser produces a highly coherent, highly directional, and nearly monochromatic beam of light, by stimulating atoms or molecules to emit light at particular wavelengths and amplifies that light, typically producing a very narrow beam of radiation.

WAVELENGTH Electromagnetic energy is transmitted in the form of a sinusoidal wave. The wavelength is the physical distance covered by one cycle of this wave.

Definitions adapted from
Oxford American English Dictionary
Encyclopaedia Britannica

Relevant reading

Brandenburg, B., Zhuang, X. (2007) Virus trafficking – learning from single-virus tracking. *Nature Reviews Microbiology* **5**: 197–208

Goldys, E. M. (2009) *Fluorescence Applications in Biotechnology and Life Sciences*. Wiley-Blackwell, New York

Lichtman, J. W., Conchello, J. A. (2005) Fluorescence microscopy. *Nature Methods* **2**: 910–919

Keats, J. (2005) The deadly art of virus cinema. *Wired* **13.8** http://archive.wired.com/wired/archive/13.08/molecular.html

Wouters, F. S. (2006) The physics and biology of fluorescence microscopy in the life sciences. *Contemporary Physics* **47**: 239–255

Photo credits

p 35   Artist conception of a virus © Bokononist – Fotolia
p 37   Artistic impression of a virus crossing a cell membrane, courtesy of Feng Zhang, Massachusetts Institute of Technology (2004)
p 37   Sequence of virus entry © 2004 Michael Rust, Melike Lakadamyali, Feng Zhang, and Xiaowei Zhuang. Courtesy of *Nature Structure and Molecular Biology Journal*

OPTOGENETICS

And other neuronal firelighters

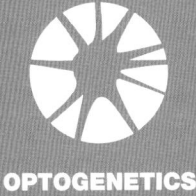

OPTOGENETICS

OPTOGENETICS

He had no acquaintances at the party, and was not really an extrovert. The food was not tasty, the music insipid, and thus – forced to stay by external circumstances, and not wanting to give the impression of being idle – he turned to the single occupation he found worth trying: looking at people's behavior. How they chose to dress, what they chose to eat, where they chose to sit. It was entertaining – just entertaining, he was aware he would probably miss in all his conclusions – to try to discover their motivations, the root of their choices. Well… as if such thing as having a choice ever existed. He smiled in amusement thinking about Buridan's ass.

Buridan's ass is a thought experiment, dating back to medieval times, devised precisely to discuss whether a choice is always based on a rational decision. It presents an ass, equally thirsty and hungry, and equally distanced from a pile of hay and a bucket of water. Since the animal cannot find a reason to prefer one option to the other, unable to take action, he dies, of both hunger and thirst. The thought experiment, that exposes the complex nature of free will by reducing its assumption to an absurd situation, is framed in a (perhaps endless) discussion that for centuries was the battleground of philosophers, now joined by neuroscientists.

For an experimental neuroscientist, analyzing behavior is a phenomenal task. The patterns behind choices and actions are woven up from tens of millions of interconnected neurons, whose firing activity takes place in a millisecond timescale. To understand in full any of those events, to lay any causal connection between them, it is necessary to have certain control of the neuronal activity, exciting and inhibiting it on demand. An example of an external tool for such control is the use of electrodes, but those may cause unintentional firing in the surroundings of the targeted neuron. Luckily for researchers, there exist algae and bacteria that find in sunlight the reason to excite or inhibit their behavior.

Such microorganisms contain light-sensitive proteins responsible for the regulation of basic survival behaviors. The motor response of some bacteria, for example, activates when illuminated to take the bacteria away from the source of light. By inserting the genes conferring this light-responsiveness into specific neurons, it is possible to control their electrical activity patterns by shedding on them beams of light of the right color. This is the basis of optogenetics.[1]

If light can trigger and inhibit neuronal activity, it can be used to establish a link between a choice of action and its cause. In such respect, optogenetics has already provided insights about the nature of certain mental disorders, such as Parkinson's disease (in animal models). Furthermore, in the labyrinth of connections that constitute each mental process, the ability to illuminate the unexplored paths stands as one of the most powerful means, not only to study the brain, but also to overcome certain impairments or barriers, laid perhaps by the lazy attitude of some neuron that chose not to fire.

[1] It should be mentioned that the genetic engineering of neurons is not the only approach to optically control their firing activity. The emerging field of optopharmacology, based on the synthesis of light-sensitive molecules, has already described interesting applications in pain control in animal models without genetic modification.

Light

may stimulate the choice of new options

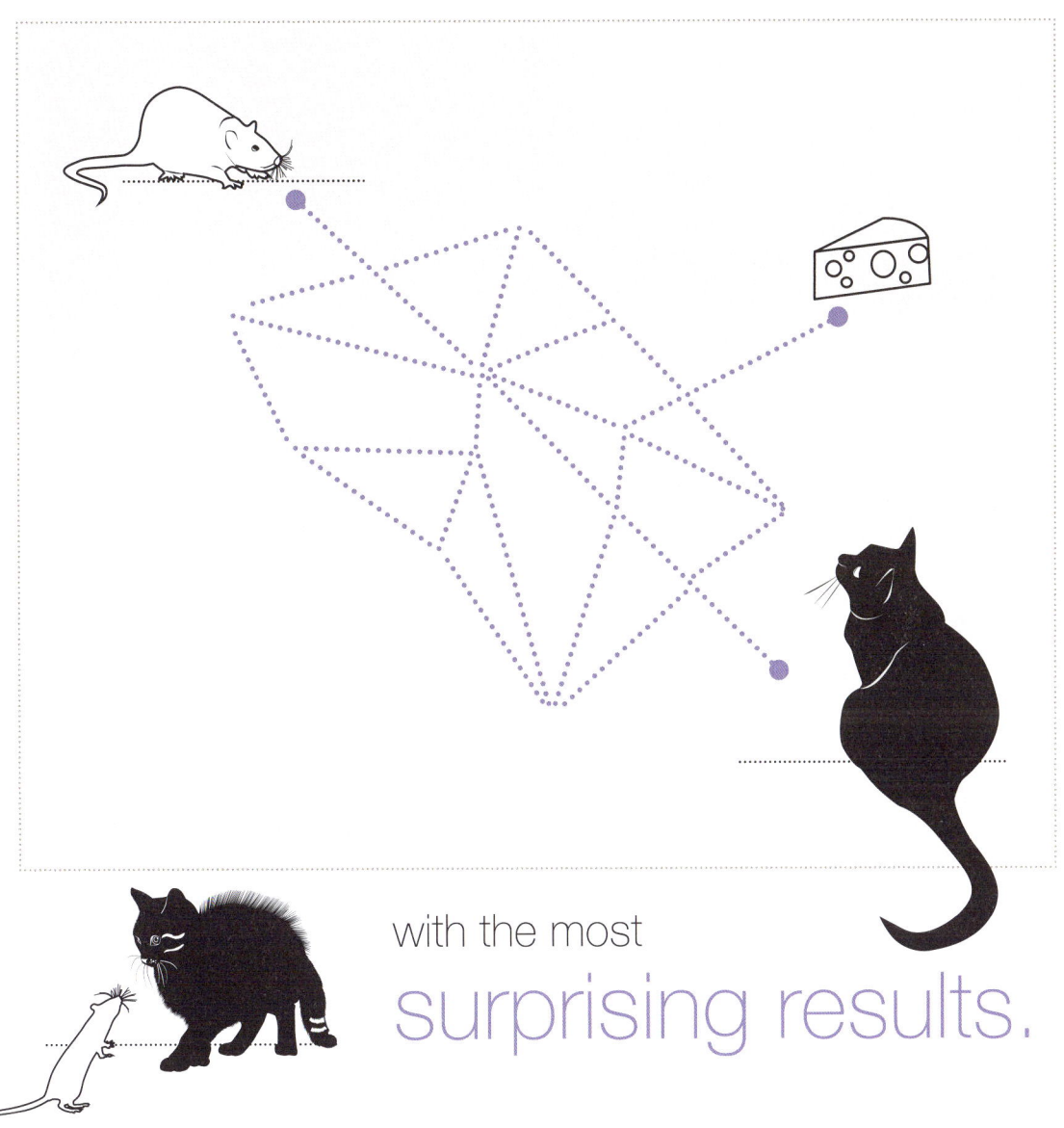

with the most
surprising results.

OPTOGENETICS

And other neuronal firelighters

Neurons are nerve cells. They communicate through electrical impulses, called action potentials, caused by the flow of ions in and out of the cell membrane. The main ions involved are sodium, potassium, and calcium (positively charged) and chloride (negatively charged). When a neuron is at rest, the cell membrane keeps the inner of the cell at a negative electric potential relative to the outside. An external stimulus may cause the sodium channels at the membrane to open. Because sodium ions have a higher concentration outside cells, they enter the neuron, causing a cascade of electric currents that ends in the firing of an action potential. After that, the membrane potential is re-established and the neuron comes back to rest.

Some types of **opsin** proteins, present in certain bacteria and algae, when activated by light are able to regulate this flow of ions across the membrane. For instance, NpHR halorhodopsin pumps in negative chloride ions in response to yellow light (**wavelength** 589 nm) – thus inhibiting the firing – and ChR2 channelrhodopsin allows positive sodium ions to pass in response to blue light (wavelength 470 nm), inducing the firing. In optogenetics, neurons are made to genetically express these proteins in the membrane by infecting them with viruses into which the opsin genes have been inserted. When light of a specific color hits an engineered neuron, the corresponding ion channels open, thus activating or deactivating the neuron firing. Light sources used in optogenetics include mercury and xenon lamps, **light-emitting diodes**, scanning lasers, or femtosecond **lasers**.

Optical manipulation with light, including optogenetics and optopharmacology, benefits from a high resolution both temporally (milliseconds) and spatially (micrometers or less) while being minimally invasive. Optopharmacology mainly differs from optogenetics in that the latter is based on the search and genetic engineering of opsin clones, while the former uses chemical synthesis to obtain the light-sensitive molecules, which have drug-like properties in certain cases.

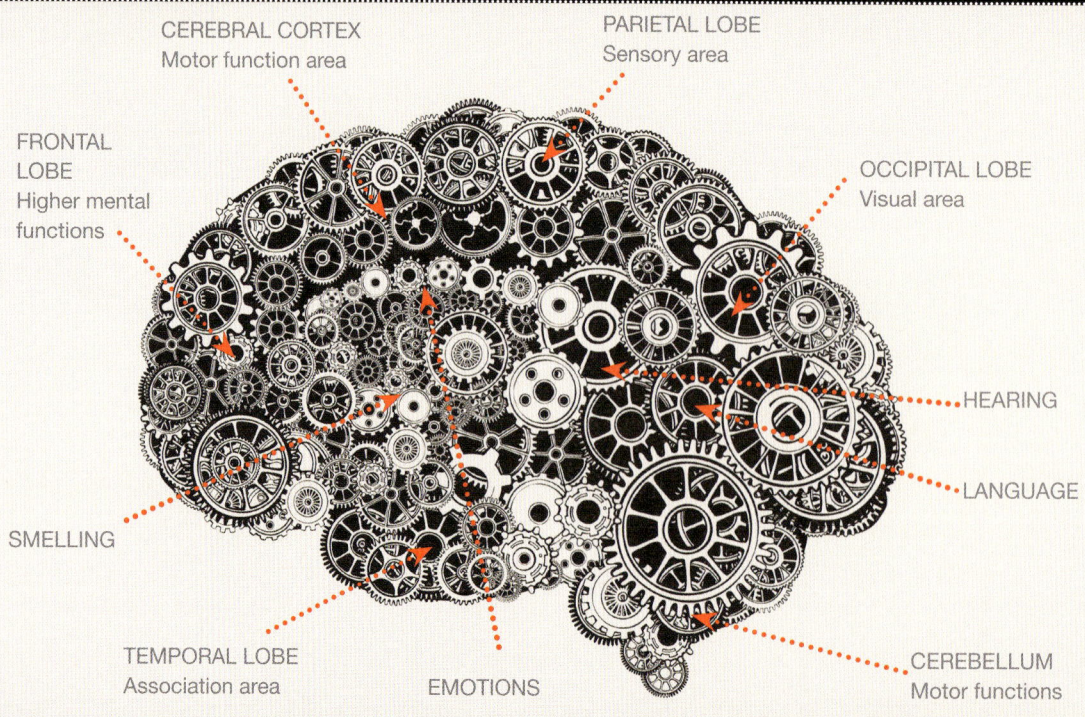

CEREBRAL CORTEX
Motor function area

PARIETAL LOBE
Sensory area

FRONTAL
LOBE
Higher mental
functions

OCCIPITAL LOBE
Visual area

HEARING

LANGUAGE

SMELLING

TEMPORAL LOBE
Association area

EMOTIONS

CEREBELLUM
Motor functions

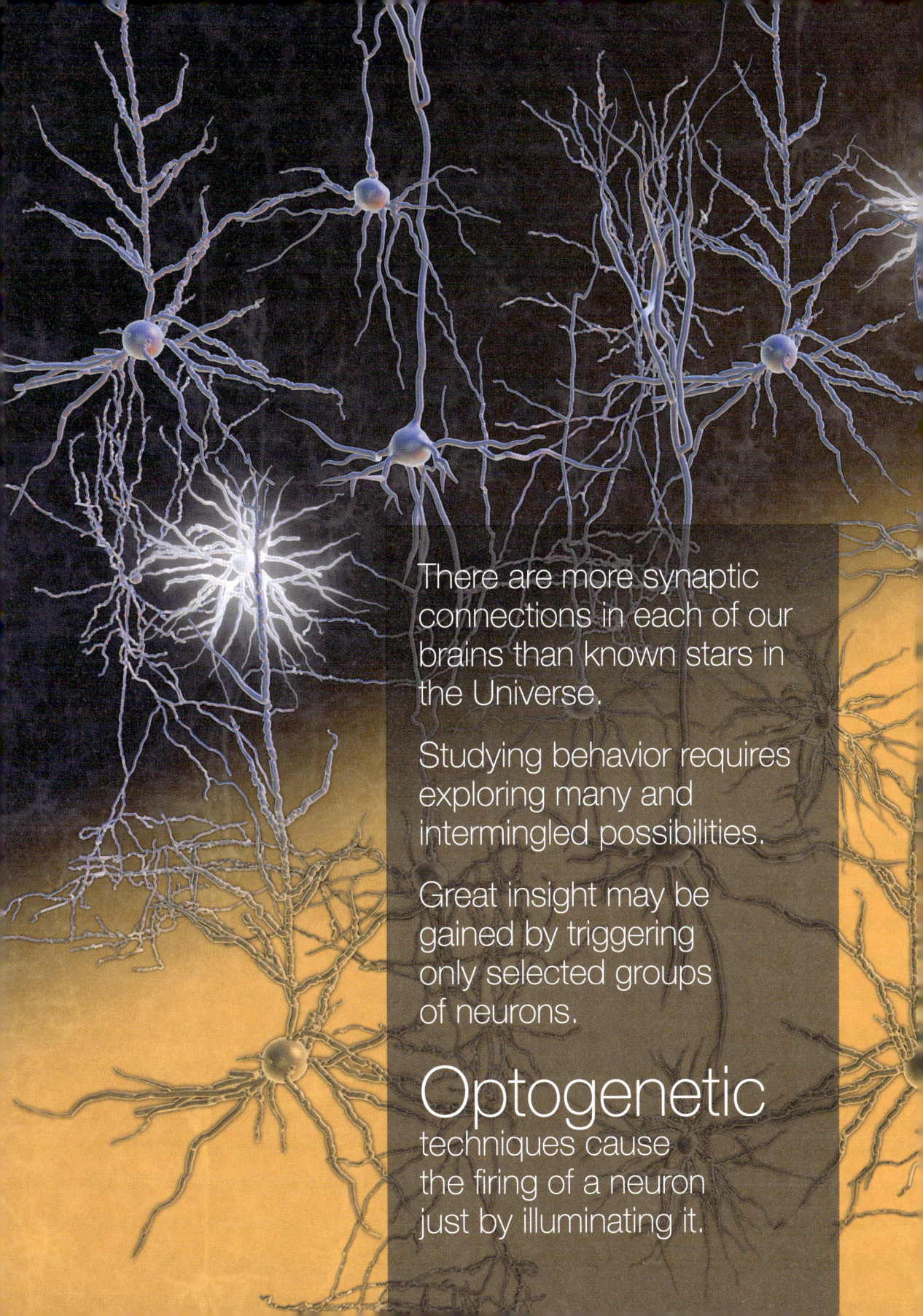

There are more synaptic connections in each of our brains than known stars in the Universe.

Studying behavior requires exploring many and intermingled possibilities.

Great insight may be gained by triggering only selected groups of neurons.

Optogenetic techniques cause the firing of a neuron just by illuminating it.

Scientific and technical advisors

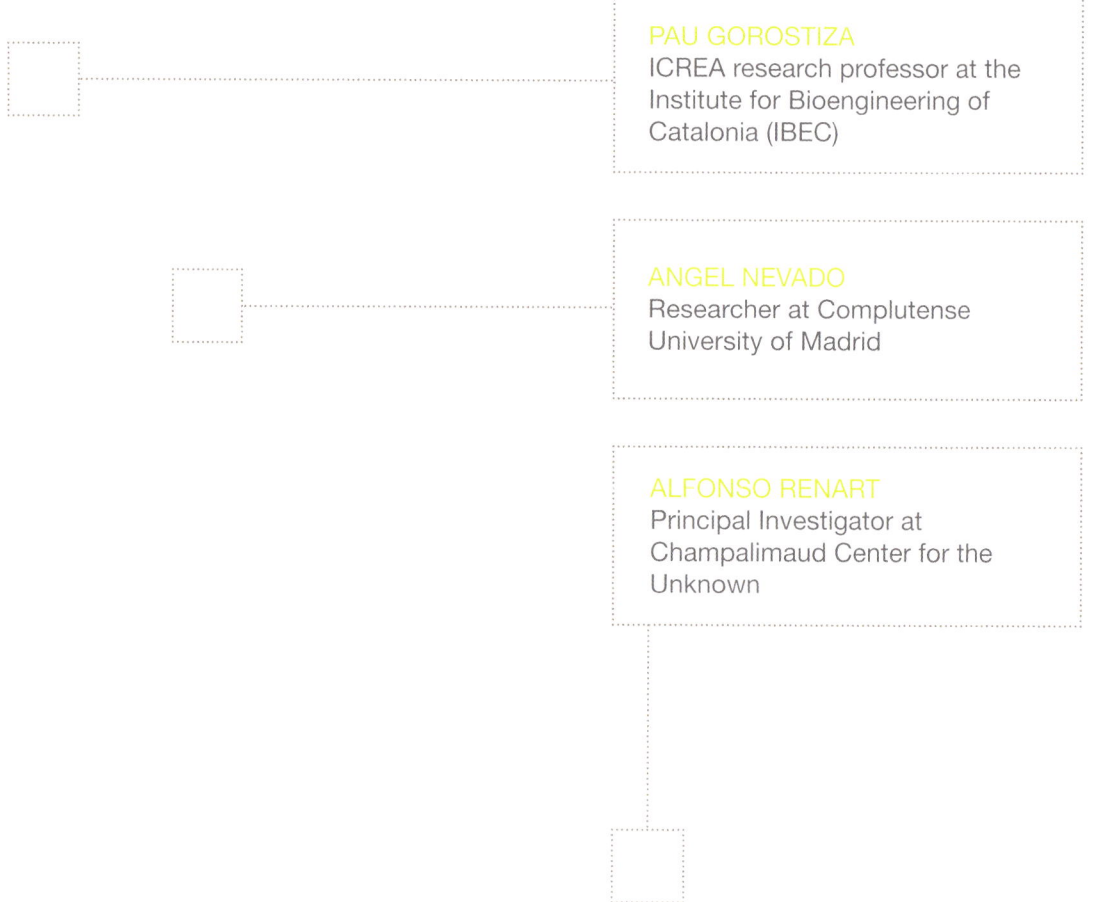

PAU GOROSTIZA
ICREA research professor at the
Institute for Bioengineering of
Catalonia (IBEC)

ANGEL NEVADO
Researcher at Complutense
University of Madrid

ALFONSO RENART
Principal Investigator at
Champalimaud Center for the
Unknown

Glossary

LASER   Acronym for Light Amplification by Stimulated Emission of Radiation. A laser produces a highly coherent, highly directional, and nearly monochromatic beam of light, by stimulating atoms or molecules to emit light at particular wavelengths and amplifies that light, typically producing a very narrow beam of radiation.

LED   Light-emitting diode. A semiconductor device which emits incoherent light when a voltage is applied. Efficiency continues to rise making an effective source of illumination for a huge number of applications.

OPSIN   Opsins are a group of proteins at the molecular basis of various light-sensing systems, including eyesight, circadian rhythms, and a type of photosynthesis. All opsin proteins are embedded in cell membranes, crossing the membrane seven times.

WAVELENGTH   Electromagnetic energy is transmitted in the form of a sinusoidal wave. The wavelength is the physical distance covered by one cycle of this wave.

Definitions adapted from
Oxford American English Dictionary
Encyclopaedia Britannica

Relevant reading

Boyden, E. S., Zhang, F., Bamberg, E., Nagel, G., Deisseroth, K. (2005) Millisecond-timescale, genetically targeted optical control of neural activity. *Nature Neuroscience* **8**: 1263–1268

Deisseroth, K. (2010) Controlling the brain with light. *Scientific American* **303**: 48–55

Gorostiza, P., Isacoff, E. Y. (2008) Optical switches for remote and noninvasive control of cell signaling. *Science* **322**: 395–399

Reiner, A., Isacoff, E. Y. (2013) The Brain Prize 2013: The optogenetics revolution. *Trends Neuroscience* **36(10)**: 557–560. Available at http://www.ncbi.nlm.nih.gov/pubmed/24054067

Photo credits

p 44   Cogs in the shape of a human brain © RYGER – Shutterstock

p 45   Artist conception of illuminated neurons, courtesy of Ed Boyden, Sputnik Animation, and the McGovern Institute for Brain Research at MIT

FAST

What a difference an attosecond makes

Time is measured by changes, e.g. heartbeats count instants, moving shadows over a landscape mark the hours, and the births and deaths of our loved ones evidence the passing of the years. Periods longer than a lifetime are increasingly difficult to properly accommodate in our mental space. Indeed, while we describe art movements in terms of centuries, and millennia might give a framework to the duration of a civilization, a block of hundreds of thousands of years lies beyond the magnitude of our imagination. We need tangible changes to understand (how much) time passes. For this reason, when a paleontologist explains that 3.5 billion years elapsed between the first signs of life on Earth and the appearance of the *Homo sapiens*, we allow our mind to open to a new unit of time.

Similarly, it is difficult to grasp time blocks that are shorter than a heartbeat. Anything happening in less than a thousandth of a second falls under the wide label of "very fast." But, how many degrees of "fast" are there? The flapping of a bee's wings, the scattering of the pieces of an exploding balloon, the functioning of modern chips, each of these processes is a thousand times faster than the previous, but all are equally indiscernible to our eye–brain system, which is limited to less than a few dozen images per second. However, there is a way of introducing ourselves as natural observers of the super-fast: splitting the whole process into still pictures. Each shot needs to record the entrance of light during a short-enough lapse of time, otherwise we would get a blurred image. Shots, moreover, should come up at a tuned pace, otherwise we would miss some relevant steps while the light is off. A stroboscopic lamp, for instance, with flashes of microseconds shooting at intervals of hundredths of a second, offers a clear sequence of the trajectory of an object as fast as a flying bullet.

The challenge is thus to produce a slow-motion movie of the fastest processes in Nature: the electronic motion determining the minute details of chemical reactions, of the order of a billionth of a billionth of a second (what we call an attosecond). This would detail with ultimate precision the steps in which atoms and molecules combine to form other substances, a process of fundamental importance, both for basic and applied research. But the illumination crew in charge of filming such a sequence would have to operate flashes of light of attoseconds themselves! This interesting conundrum is solved by X-ray laser pulses, a source of light going exactly where we want, precisely when we want, and with just the amount of energy we want.

Everything
is continuously
changing.

Registering a change
implies finding a system that
continuously changes itself
at an equal or faster pace.

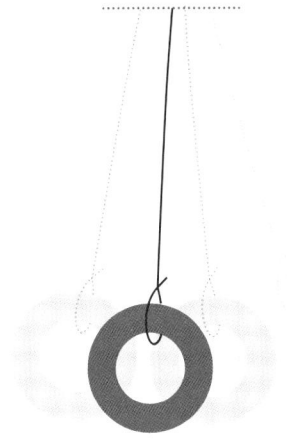

So...

...how are
we to measure the
fastest events in the
Universe?

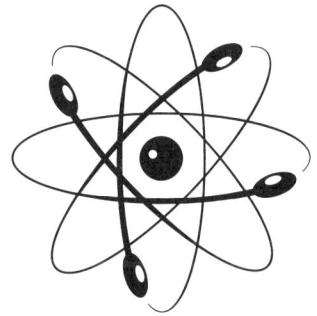

FAST

What a difference an attosecond makes

The image we see at the right uses short flashes of light to reproduce a clear sequence of the trajectory of a diver. The length and pace of those flashes depend on the physical process that generates them: in this case an intense electric discharge on a tube containing ionized xenon. Harold E. Edgerton, a brilliant professor of electrical engineering at MIT, and a passionate photographer, started designing those strobo-scopic lamps in the 1930s.

Similarly, the dynamic behavior of electrons within a chemical reaction can be resolved by "flashes" of light, in this case on an attosecond timescale. These ultrashort pulses of light are generated by very intense (more than 10^{14} watts per square centimeter) and very short (femto-second) **laser** pulses directed over a chamber containing a small quantity of helium. When an intense electric field hits an atom, it accelerates an electron first away and then back to the parent ion. Returning to the ion, the electron recombines, creating a **soft X-ray** burst of light at an attosecond timescale.

Especially interesting for viewing living speci-mens is producing these flashes in the so-called "water window," electromagnetic radiation in the **soft X-ray** region with **wavelengths** between 2.3 nm and 4.4 nm. Since water is transparent to these wavelengths, such flashes would allow imaging of the ultrafast dynamics of biochemical reactions *in vivo*.

Control over electron dynamics can improve all processes where movement of electrons is involved. Applications can be found in health – reducing undesirable by-products in synthesis of drugs, energy – improving the efficiency of fuels, and information – and upgrading performance of computer chips, to name a few.

It has been about 10^{18} seconds since the beginning of the Universe.

Pause for a moment to grasp that number.

Now divide a second in 10^{18} parts.
You get an attosecond.

Pause for another moment to grasp that number.

An attosecond is the timescale for the movement of electrons.

It's a tiny instant of time, unimaginable for us, and yet, the world does change from one atto-second to the next.

An ultrafast laser can witness that change: pulses of 200 attoseconds allow filming of the intimate steps of chemical reactions by accurately tracing the dynamics of the electrons involved.

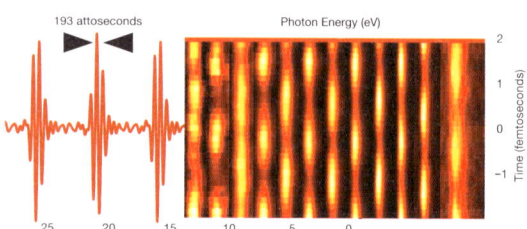

Direct observation of the electric field of light.
Experimental data showing the photoelectron spectrum in an ionized gas, used to measure pulses of about 200 attoseconds.

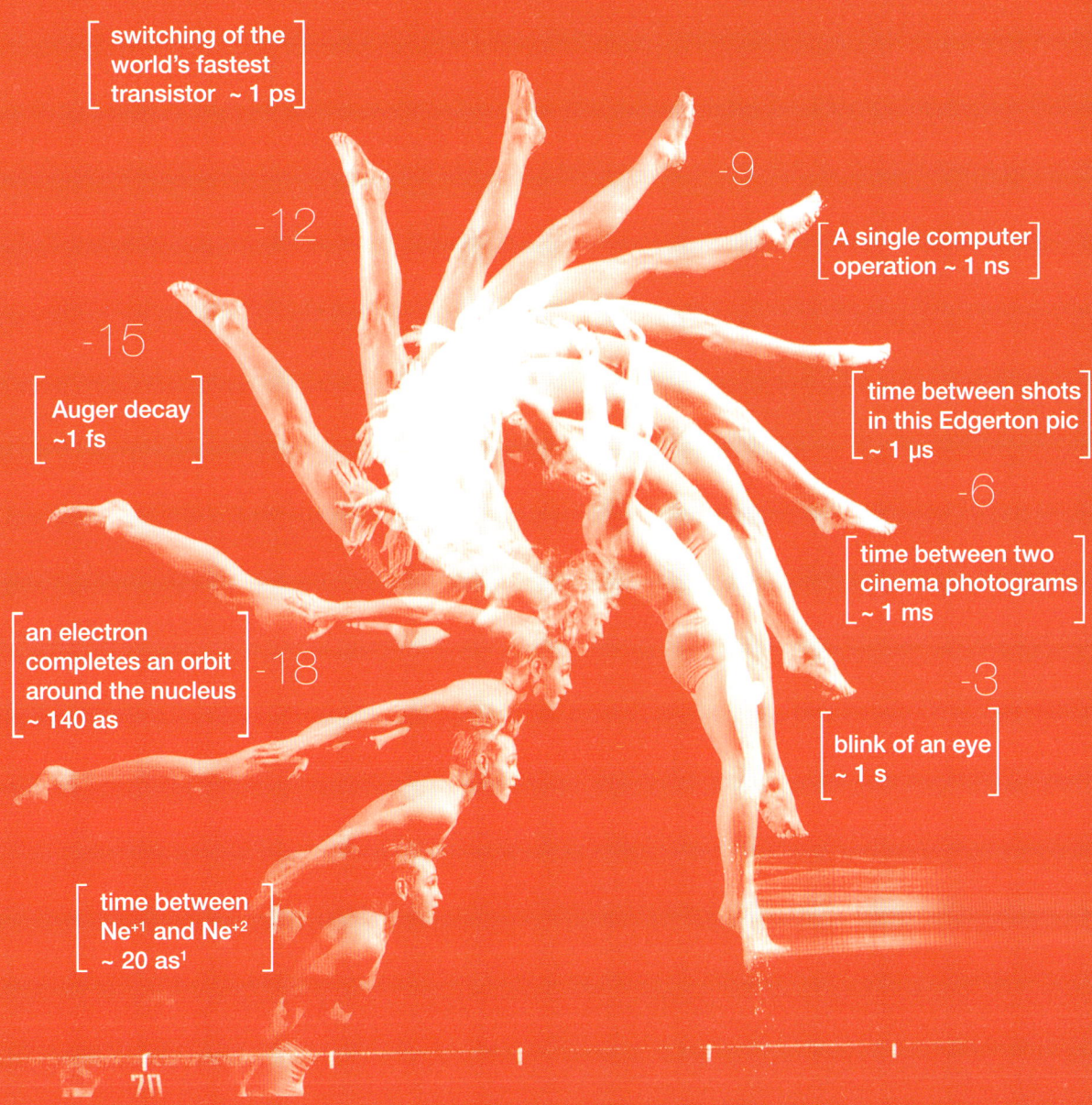

switching of the
world's fastest
transistor ~ 1 ps

-12

-9

A single computer
operation ~ 1 ns

-15

Auger decay
~1 fs

time between shots
in this Edgerton pic
~ 1 μs

-6

time between two
cinema photograms
~ 1 ms

an electron
completes an orbit
around the nucleus
~ 140 as

-18

-3

blink of an eye
~ 1 s

time between
Ne^{+1} and Ne^{+2}
~ 20 as[1]

[1] Ne^{+1} stands for an atom of neon that has lost one electron (an ion). Ne^{+2} is the second ionization

Scientific and technical advisors

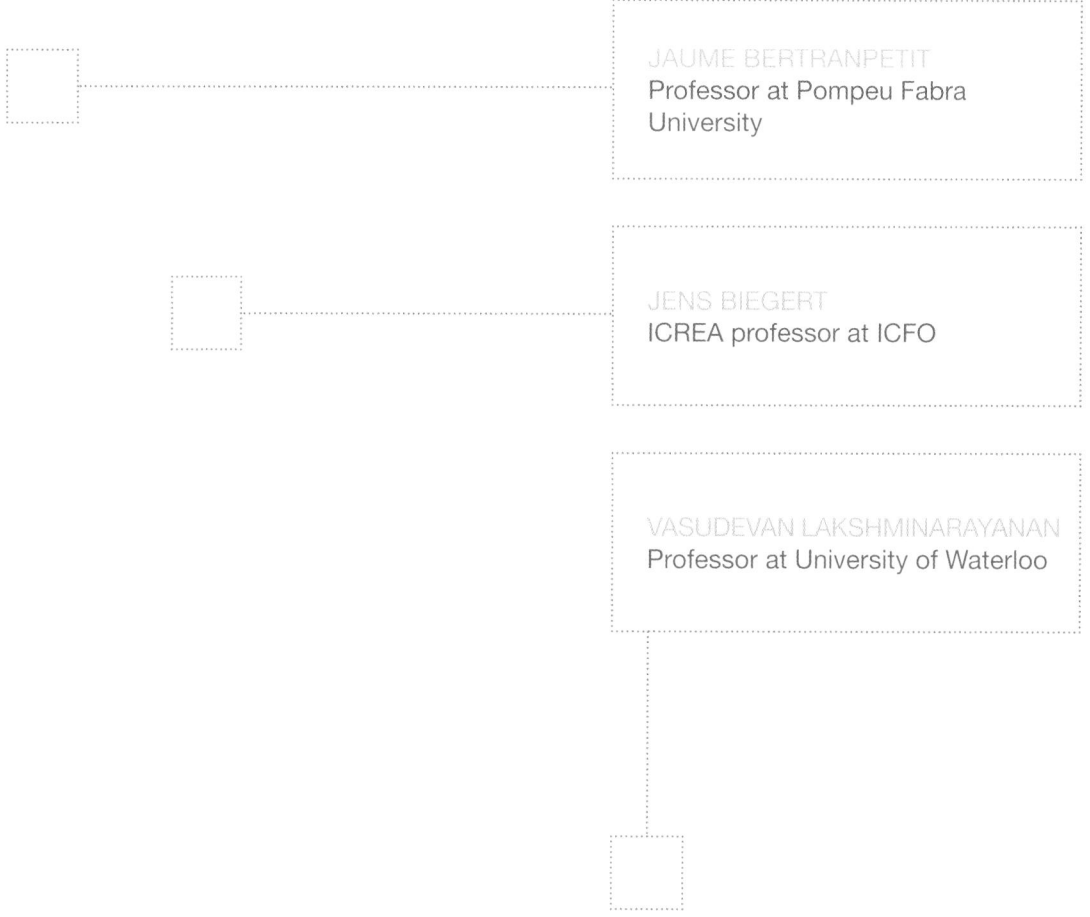

JAUME BERTRANPETIT
Professor at Pompeu Fabra
University

JENS BIEGERT
ICREA professor at ICFO

VASUDEVAN LAKSHMINARAYANAN
Professor at University of Waterloo

Glossary

AUGER DECAY   The spontaneous emission of an electron instead of a photon in the relaxation of an excited atom.

LASER   Acronym for Light Amplification by Stimulated Emission of Radiation. A laser produces a highly coherent, highly directional, and nearly monochromatic beam of light, by stimulating atoms or molecules to emit light at particular wavelengths and amplifies that light, typically producing a very narrow beam of radiation.

ULTRAVIOLET   The portion of the electromagnetic spectrum extending from the violet end of the visible region to the X-ray range. The ultraviolet range is usually divided into three regions: near-ultraviolet, from 400 nm to 200 nm; far-ultraviolet, from 200 nm to 30 nm; and extreme-ultraviolet, from 30 nm to 10 nm.

WAVELENGTH   Electromagnetic energy is transmitted in the form of a sinusoidal wave. The wavelength is the physical distance covered by one cycle of such a wave.

X-RAY   The portion of the electromagnetic spectrum extending from extreme-ultraviolet to the gamma-ray range. The corresponding wavelengths span from about 10 nm to 10 pm.

Definitions adapted from
Oxford American English Dictionary
Encyclopaedia Britannica

Relevant reading

Repneck, J. (2003) *The Man who Found Time.* Perseus, Cambridge, MA

Clegg, B. (2007) *The Man who Stopped Time.* Joseph Henry Press, Washington

Eastman House Editors (1964) *Seeing the Unseen: The Life and Work of Harold Edgerton*. Eastman House, Cambridge, MA

Landes, D. S. (1983) *Revolution in Time*. Belknap Press of Harvard University Press, Cambridge, MA

Smolin, L. (2013) *Time Reborn: From the Crisis in Physics to the Future of the Universe*. Houghton Mifflin Harcourt, Boston

Zewail, A. H. (1999) Nobel Lecture. http://www.nobelprize.org/nobel_prizes/chemistry/laureates/1999

Photo credits

p 52 Direct observation of the electric field of light © 2010 ICFO, Jens Biegert's group

p 53 Multiflash picture of diver, Burt West. Picture by Harold Edgerton © 2010 MIT, courtesy of MIT Museum

RANDOM WALKS

The odyssey of photons

RANDOMWALKS

Photons travel in vacuum following straight lines. Their eight-minute flight from the Sun to the surface of the Earth is a peaceful journey almost all the way through. Hundreds of kilometers before reaching the ground, though, they start facing several layers of obstacles. The upper atmosphere reflects the most energetic photons back into space. From the remaining photons, a portion is absorbed by some of the tens of molecules identified across the following layers of the atmosphere. The rest reach their destination on time, except for a few that may be still delayed in the next-to-last stop. Much like Ulysses in Homer's *Odyssey*, who spent ten years bouncing between islands and coasts to cross the few kilometers of Aegean waters from Troy to Ithaca, these last few photons undergo a random walk from molecule to molecule in the concluding layers of the atmosphere. Eventually they land, continuously and from all directions, filling with bright blue the eyes that look at the sky.

The above-described adventure of photons results in an inexhaustible source of information about the composition and behavior of atmospheric gases. The HITRAN (high resolution transmission molecular absorption) spectroscopic database, maintained and developed at the Harvard–Smithsonian Center for Astrophysics in Cambridge, Massachusetts, archives the spectral parameters (the identity card) of the molecules absorbing and scattering light in the sky. People working at HITRAN like to consider it as the genome project for molecules and their isotopic variants. HITRAN provides data to academia and

industry around the world, aids in the remote sensing of other planets' atmospheres, and helps track the presence and concentration of polluting agents and allergens in the Earth's air.

Such analysis of absorption and scattering of light may be particularized for the detailed study of any diffusive medium. The specific study of the diffusion of light in human tissues is a resourceful method in medical sciences, where physiological features can be assessed non-invasively and often in real time. By shining light on blood vessels and quantifying how much of it is absorbed, it is possible to keep track of the amount of oxygen present in the blood, since the oxygenated and deoxygenated forms of hemoglobin – the protein of the red blood cells in charge of carrying oxygen around the organism – absorb light in discernible ways. A main direct application is the mapping in real time of neural activity in critical regions of the cerebral cortex. Scattering of light, on the other hand, is characterized by the length of the free path that photons travel before encountering an obstacle. Since any variation of such a path may be mostly associated with irregularities in the medium, the diffusion of photons through tissue can help in the detection of early tumors or abnormal blood flows.

Photons do not travel in a straight line through human tissue. While their random walks prevent the acquisition of sharp images of the inner body taken from outside, priceless information can be retrieved from photons that wander through the red blood rivers of human bodies.[1]

---

[1] There are other optical modalities, like OCT or acousto-optics, for clinical imaging, but here we focus on pure diffuse optics.

Random walks in blue and red

THE LAESTRYGONES

CIRCE

CIRCE

TO PILLARS OF HERACLES

TROY

THE CYCLOPES

ITHACA

CHARYBDIS

Ulysses
did not make
it home in a
straight line.

CALYPSO

LOTUS EATERS

THE ROUTE OF ULYSSES

He spent ten years wandering the
sea, driven by the waves or held
captive by fickle goddesses. And
he learned a lot in the process...

...just like photons.

SOURCE

DETECTOR

Absorption
and scattering
of photons in
a medium give
important information
about it.

RANDOM WALKS

The odyssey of photons

When light enters human tissue, it is either absorbed by specific molecules (hemoglobin, water, lipids) or scattered by heterogeneities in the tissue (mitochondria, cell membranes, red blood cells). When the amount of absorption sits well under the amount of scattering, random walks or migration of photons are possible. In this regime, called photon diffusion, light can penetrate several centimeters into the tissue.

Medical diffuse optics is a set of techniques based on the photon diffusion model to infer physiological properties from optical properties (absorption and scattering). The main medical applications to date are breast cancer imaging and monitoring of the brain activity and well-being. All these techniques work with inexpensive, fast, portable, and non-invasive instrumentation.

The light source in diffuse optics is either coherent (laser) or non-coherent (light-emitting diode, lamp) in the red or near-infrared, spanning the optical window between wavelengths of 650 and 1000 nm. The diffusion model separates the effects of absorption from scattering, so that experimenters can directly measure the concentration of specific molecules using their well-known absorption spectra (see physiological window below), and separately infer the presence of irregularities. Furthermore, when laser light is used, laser speckles can also be detected, which yields real-time measurements of blood flow.

Applications in personalized cancer treatment derive from the increased vascularity and rapid cell division in cancerous tumors, which allow their characterization by optical means. The non-invasive study of the dynamics of blood in the brain vessels has applications in neurosciences and neurology, allowing for bed-side monitoring of patients in intensive care units.

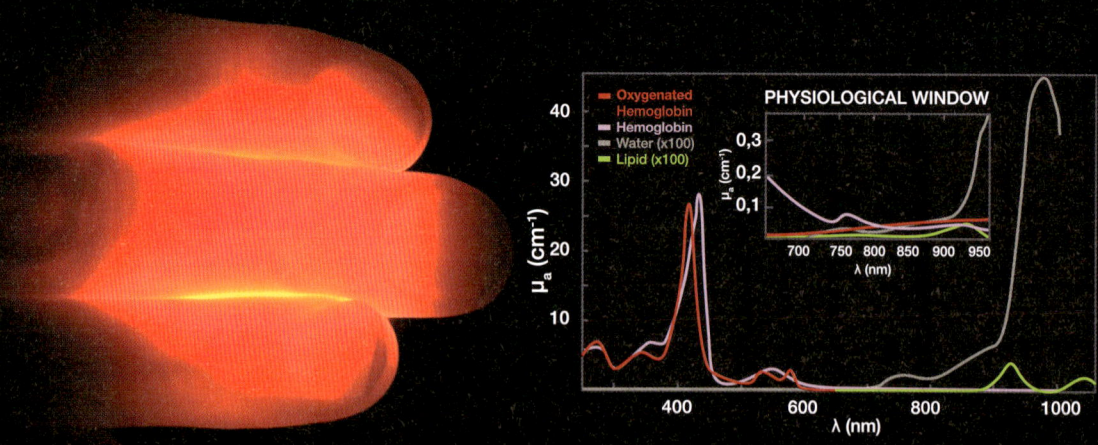

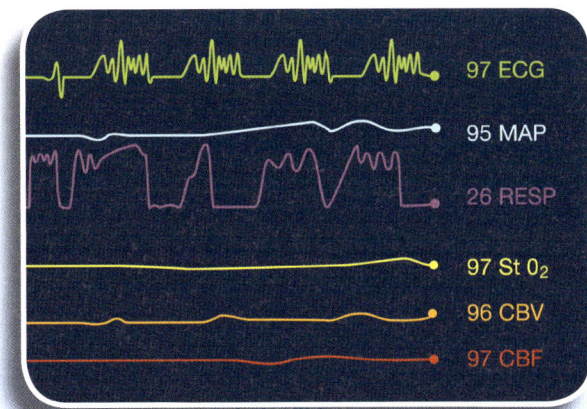

97 ECG

95 MAP

26 RESP

97 St 0₂ — Blood Oxygen Saturation

96 CBV — Cerebral Blood Volume

97 CBF — Cerebral Blood Flow

When photons travel through a medium, different components absorb light at specific frequencies, so that particular absorption patterns inform us about the amount of certain components in blood.

A device based on diffuse optics is able to non-invasively monitor blood flow and oxygenation, leading to individualized therapies and care.

Analyzing the effects of collisions in the trajectories of the photons provides critical information about blood flow.

Scientific and technical advisors

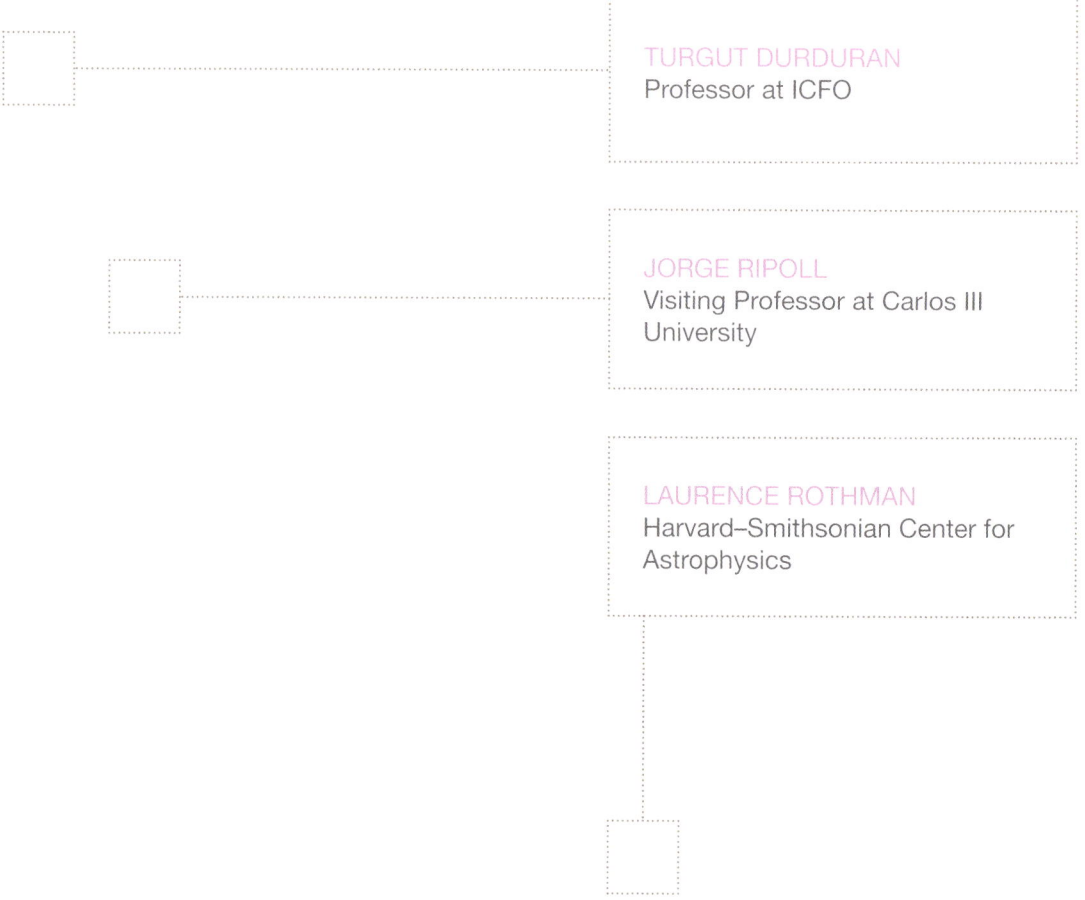

TURGUT DURDURAN
Professor at ICFO

JORGE RIPOLL
Visiting Professor at Carlos III University

LAURENCE ROTHMAN
Harvard–Smithsonian Center for Astrophysics

Glossary

ABSORPTION SPECTRUM — A pattern of dark lines observed in the spectrum of light when it passes through a particular absorbing medium, corresponding to the missing frequencies that have been absorbed. Such a pattern is unique and can be used to identify the medium.

INFRARED — The portion of the electromagnetic spectrum extending from the red end of the visible region to the microwave range. The infrared range is usually divided into three regions: near-infrared, from 700 nm to 2 μm; medium-infrared, from 2 μm to 4 μm; and far-infrared, from 4 μm to 1 mm.

LASER — Acronym for Light Amplification by Stimulated Emission of Radiation. A laser produces a highly coherent, highly directional, and nearly monochromatic beam of light, by stimulating atoms or molecules to emit light at particular wavelengths and amplifies that light, typically producing a very narrow beam of radiation.

(LASER) SPECKLES — Area of coherence produced due to interference pattern of light that travelled different pathlengths.

LED — Light-emitting diode. A semiconductor device that emits incoherent light when a voltage is applied. Efficiency continues to rise making an effective source of illumination for a huge number of applications.

WAVELENGTH — Electromagnetic energy is transmitted in the form of a sinusoidal wave. The wavelength is the physical distance covered by one cycle of this wave.

Definitions adapted from
Oxford American English Dictionary
Encyclopaedia Britannica

Relevant reading

Durduran, T., Choe, R., Baker, W. B., Yodh, A. G. (2010) Diffuse optics for tissue monitoring and tomography. *Reports on Progress in Physics* **73**: 076701

Jaques, S. L., Pogue, B. W. (2008) Tutorial on diffuse light transport. *Journal of Biomedical Optics* **13(4)**: 041302

Ripoll, J. (2012) *Principles of Diffuse Light Propagation*. World Scientific, Singapore

Yodh, A., Chance, B. (1995) Spectroscopy and imaging with diffusing light. *Physics Today* **48**: 34–40

The Hitran Database – http://www.cfa.harvard.edu/hitran/

Photo credits

p 59   Illustration of the random walk of photons through tissue © ICFO-Digivision

p 60   Light through human tissue © ICFO

p 61   Head model © padavan – Fotolia

DISPLAYS

The great escape of images

DISPLAYS

DISPLAYS

Vision strongly influences our perception. Examples are the McGurk effect, an auditory illusion where the visual component of speech clearly influences what a person hears; or the famous experiments conducted in 2001 by Brochet and Dubourdieu, involving 57 wine experts, in which every one of them failed to notice that it was actually a white wine with an odorless red dye what they were asked to rate as a red wine.

Further off senses, images enjoy the power of acting as a kind invitation, a call to mind, of more abstract ideas. In geometry, the most visual part of mathematics, the flow of reasoning can generally be followed by looking at images. Elsewhere in the sciences and the arts, production of explanatory, enlightning, summarizing pictures are often the basis of a successfull spread.

Currently, technology is working in optimal displays to make the power of vision even stronger.

Displays are indeed the material interface in which the invisible gets converted into the visible. They provide graphical interpretations of realities beyond the capacity of our senses. For example, most information in the non-visible regions of the spectrum, from the gamma and X-ray to the radiowave, is translated into the visible, incorporating the nanoworld and the astronomical world – from the ultra-small and the ultra-large, and from the ultra-fast and the ultra-slow – into human-scale perception.[1] Displaying artistic images of phenomena that are not encountered in the macroscopic world, such as the operation of the intra-cellular molecular machinery, opens the imagination to a whole new world.

In a connected world, where technology struggles every day to store and transmit bigger, faster loads of information, within a knowledge-based society depending on its capacity of aquiring, integrating, and distributing such information, the massive and global wiring of data might be rendered incomplete without an appropriate canvas to display it at the other side.

A good canvas does not necessarily imply a high resolution, as the power of images does not always lie in being an exact representation of reality. A good canvas needs to facilitate the assimilation of an object's specific representative lines from every relevant point of view, befitting the user's circumstances. For instance, offering interactive content for schools with restricted access to printed resources; or unrolling the full blueprint of a pharaoh's tomb in the middle of the desert; or projecting a manipulable 3D image of a human heart directly over the surgical table.

Images are already escaping from the rigid plain world of bulk computer screens. They have made it to domestic and city furniture, smartphones, tablets, smart-glasses, and many kinds of wearable devices. Display technologies based on liquid crystals, light emitting diodes, and lasers complemented by nanostructured surfaces to avoid reflections, stains, or bacteria allow images to emerge in ever more flexible, rollable, cleaner, and interactive formats.

Computers, optical fibers, wireless networks, and displays might make information travel from virtually *anywhere* to the highly visible. A present-day debate might attract more attention from the general public: should there be a limit on the kind of "invisible" information that we are prepared to visualize and on the growing pervasiveness of displays?

---

[1] A timely, thrilling example is the visualization of the cosmic microwave background radiation map (see page 10), a footprint of the first stages of the Universe that is about to deploy information about crucial aspects of its evolution.

Images...

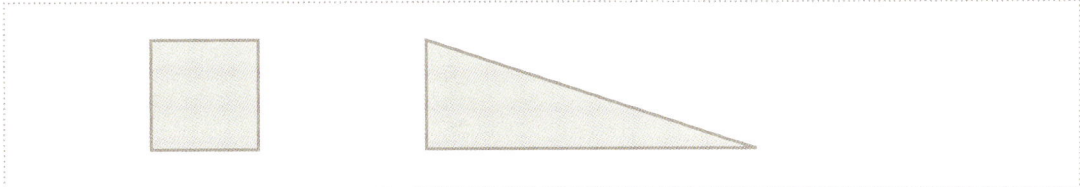

... help visualize the core ideas

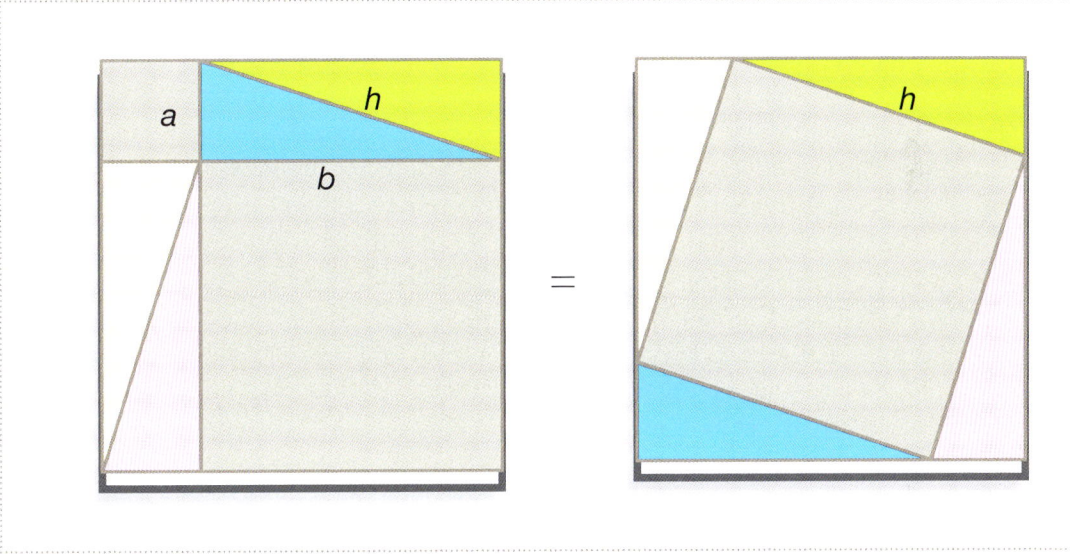

in long demanding intellectual discourses.

$$a^2 + b^2 = h^2$$

The goal is to project these images

in any possible situation to bring

ideas everywhere to be used.

DISPLAYS

The great escape of images

The field of display technologies sits at the intersection of optics, electronics, material science, and human visual perception. It is a huge field in constant expansion, pulled and pushed by research advances, industrial constraints, and market needs. We offer here a simple general description of the optical principles behind modern displays.

Some displays use backlight and include a layer of an active material to control the color and illumination of each pixel. For instance, in **liquid-crystal** displays (LCD), a liquid crystal is placed between two crossed polarizers. The liquid crystal affects the **polarization** of light, so when it is not active, no light can go through. Once the liquid crystal is activated, through voltage, it controls pixel by pixel the intensity and the color of the transmitted light, and an image is formed.

Other displays emit light themselves. For example, **LED** displays can generate a wide spectrum of colors by controlling color and intensity of each (red, blue, or green) pixel.

The above displays are usually rigid. Flexibility may be achieved with organic LEDs (**OLED**s), and possibly with graphene and graphene-like materials, which can also be transparent and rollable.

Transparent thin surfaces – with different polarization and refraction properties to project images – are the basis of wearable displays, from head-mounted displays to smart glasses and smart contact lenses. Other technologies aim at projecting the image directly onto the retina of the eye.

Most standard 3D images work by filtering for each eye a different view of the same object. Both views are then fused in the visual system of the brain to provide the 3D effect. The filtering can be done with or without glasses. Glasses-free 3D images are possible by including lenses onto the display or by means of **parallax** barrier systems. A different approach – not standard yet for displays technologies – involves 3D imaging by **diffraction** of several light beams (like a hologram). Some of these technologies employ **laser** beams, but 3D images can also be formed over a standard screen (for instance LCD) covered by special nanopatterned grooves that allow diffraction.

Any surface can become a display with an appropriate projector. A laser and a moving micromirror scan an image over opaque or transparent surfaces, or on a diffuse substance like mist or smoke to form a 3D object.

Scientific and technical advisors

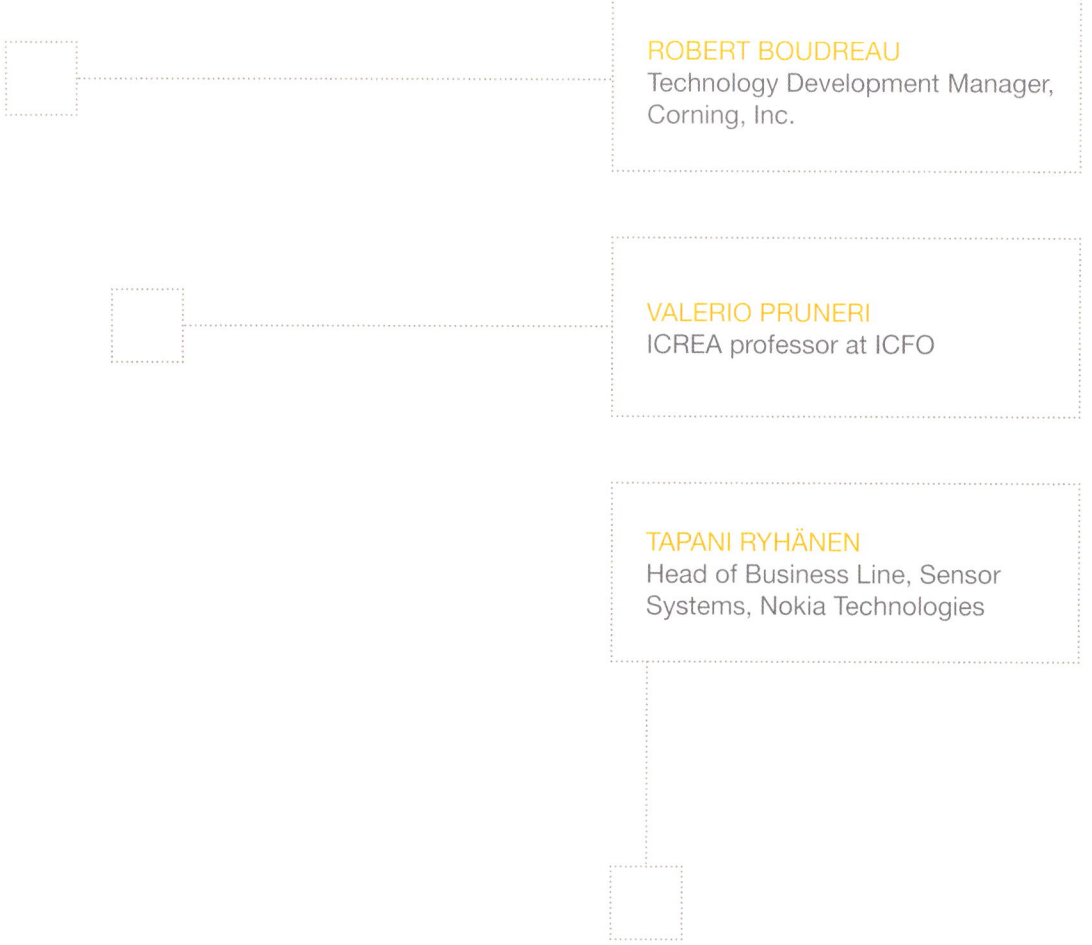

ROBERT BOUDREAU
Technology Development Manager,
Corning, Inc.

VALERIO PRUNERI
ICREA professor at ICFO

TAPANI RYHÄNEN
Head of Business Line, Sensor
Systems, Nokia Technologies

Glossary

....

DIFFRACTION As a wavefront of light passes by an opaque edge or through an opening, secondary weaker wavefronts are generated, originating at that edge. These secondary wavefronts will interfere with the primary wavefront as well as with each other to form various diffraction patterns.

LASER Acronym for Light Amplification by Stimulated Emission of Radiation. A laser produces a highly coherent, highly directional, and nearly monochromatic beam of light, by stimulating atoms or molecules to emit light at particular wavelengths and amplifies that light, typically producing a very narrow beam of radiation.

LED Light-emitting diode. A semiconductor device that emits incoherent light when a voltage is applied. Efficiency continues to rise making an effective source of illumination for a huge number of applications.

LIQUID CRYSTAL A liquid crystal is an intermediate phase where the material exhibits certain crystal-like and liquid-like properties simultaneously. Liquid crystals can change the polarization of light when a voltage is applied.

OLED Organic light-emitting diode. Light-emitting device composed of a matrix of diodes and organic light-emitting phosphors.

PARALLAX The effect whereby the position or direction of an object appears to differ when viewed from different points of view, e.g. from each eye.

POLARIZATION In a beam of electromagnetic radiation, the polarization direction is the direction of the oscillation of the electric field of light.

REFRACTION The bending of oblique incident rays as they pass from a medium having one refractive index into a medium with a different refractive index.

....

Definitions adapted from
Oxford American English Dictionary
Encyclopaedia Britannica

Relevant reading

Bradbury, R. (1967) *Fahrenheit 451*. Simon & Schuster, New York

Dourish, P., Bell, G. (2011) *Divining a Digital Future*. MIT Press, Cambridge (MA)

Fattal, D., Peng, Z., Tran, T., *et al*. (2013) A multi-directional backlight for a wide-angle, glasses-free three-dimensional display. *Nature* **495**: 348–351

Hamilton, E. (1965) *Crashing Suns*. Ace Books, New York

Mann, S., Niedzviecki, H. (2001) *Cyborg: Digital Destiny and Human Possibility in the Age of the Wearable Computer*. Doubleday, Toronto

Weiser, M. (1991) The computer for the twenty-first century. *Scientific American* **265 (3)**: 94–104

Photo credits

p 68   Optical textures exhibited by donor-acceptor stabilized columnar liquid crystals, courtesy of Professor Lee Park, Department of Chemistry, Williams College

p 69   Touch screen © peshkova – Fotolia

CONNECTED

Guided and wireless communications

CONNECTED

CONNECTED

August in a small Mediterranean village. Too hot to leave the shade of the vines and fig trees in the courtyard, the day goes by reading, chewing fruits, playing cards. Two hours after sunset, it's time to leave for the fields and listen to the crickets. A remote memory from childhood appears: somebody's grandfather is explaining a story, an old rural formula to judge the air's temperature from the frequency of the crickets' song. A sudden jolt of restlessness. Could that be true? Oh, never mind. We're lost, on holidays, in the middle of nowhere; sitting over subterranean water streams, among soft hills, under the stars. Only time comes along. Over three hours, according to the smartphone, which besides such information displays several apps to measure the air's temperature from the crickets' song. Apparently, the phone did not forget about that brief inquisitive moment a few hours ago. It's the first event of the day that happens fast.

Connected and immediate. The sign of our times involves an overwhelming flow of information from long-distance sources, primarily supported by optical fiber technology. A large part of the flow of information started leaving the saturated radio and microwave frequencies years ago. Almost all long-haul telephony and data communication today is efficiently packed into near-infrared light travelling on glass fibers. But, why light?

The very basic principle is simple. The frequency of the electromagnetic waves in the optical range is much higher than that of radio and microwaves, meaning that in a second there are more wave peaks to modulate for the encoding of information. Therefore, more ones and zeros – the alphabet of digital information – can be packed in less time.

Notwithstanding how simple the basic principle might seem, its practical implementation for the accurate transmission of information over long distances represents a serious challenge.

Alexander Graham Bell managed to send voice and other sounds encoded in a beam of sunlight. He called his invention the photophone. Until his death, he considered the prospects of the photophone to be more promising even than those of the telephone. And he has been proved right. Only technology had to make a first step in providing a source of light more controllable and efficient than the Sun. Such a source arrived with the invention of the laser. Nowadays, free-space optical transmissions are a reality, and they may even be used in inter-satellite communications. Down on Earth, light needs to be confined and guided through wires, since it is much more affected by reflection and losses on mountains, buildings, clouds, or fog. Interestingly enough, the dawn of the nineteenth century also saw the first demonstration of light guiding inside what can be considered the first "optical fiber." In 1880, Jean Daniel Colladon demonstrated how to transmit light by total reflection inside a water jet. Again, a major step from technology was needed. This time it had to provide a transmitting medium with minimum losses. Such fast, reliable fiber-based optical communications arrived only after the purification of a suitable transparent confining and transmitting medium: purified fused silica.

The story of the success of optical fibers is an example of the determination of a heterogeneous community of thinkers, what Jeff Hecht called a City of Light in his touching account of the field. It came about thanks to a constructive effort of thousands of outstanding thinkers to bring light into the inner countryside and wire the song of the crickets.

Long-haul communications were installed at radio and microwaves.

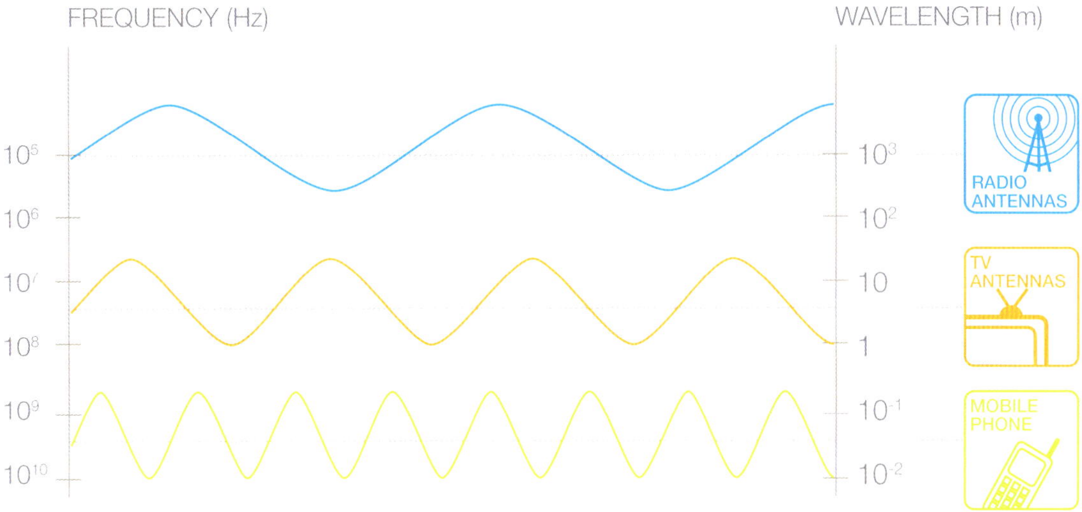

FREQUENCY (Hz) WAVELENGTH (m)

10^5 10^3 RADIO ANTENNAS

10^6 10^2

10^7 10 TV ANTENNAS

10^8 1

10^9 10^{-1} MOBILE PHONE

10^{10} 10^{-2}

They moved to optical frequencies.

Why light?

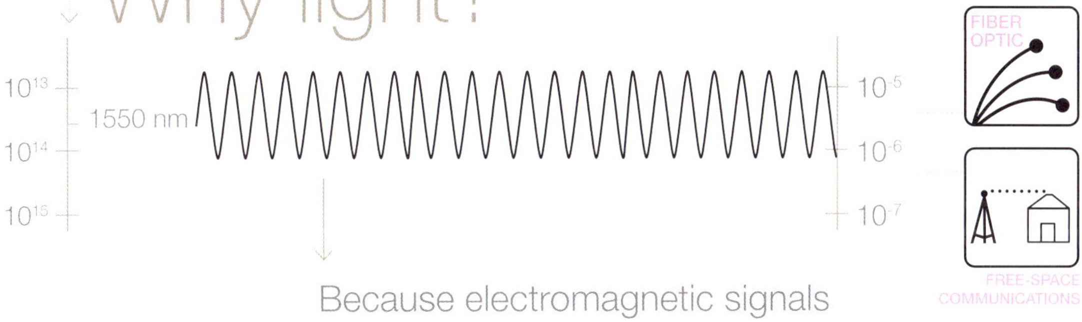

10^{13} 1550 nm 10^{-5} FIBER OPTIC

10^{14} 10^{-6}

10^{15} 10^{-7} FREE-SPACE COMMUNICATIONS

Because electromagnetic signals
with higher frequencies can pack
more information in less time.

The challenge is to transport light

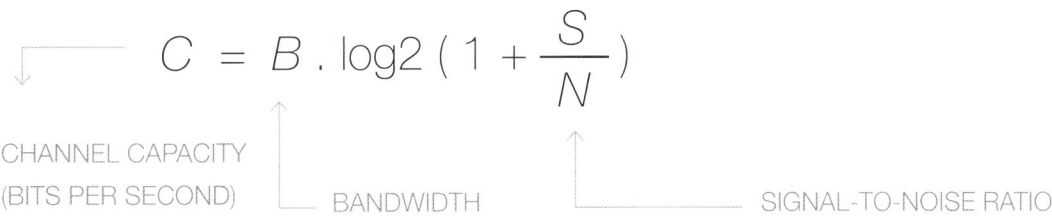

$$C = B \cdot \log2 \left(1 + \frac{S}{N} \right)$$

CHANNEL CAPACITY
(BITS PER SECOND)   BANDWIDTH   SIGNAL-TO-NOISE RATIO

CONNECTED

Guided and wireless communications

The carrying capacity of a communication channel (measured in bits per second) depends mainly on the channel **bandwidth**, which is the range of **frequencies** that the medium can transmit without significant distortion. The bandwidth of a channel is smaller than the transmitted frequencies. A broader bandwidth, which can accommodate more information, may be achieved by using higher carrier frequencies. This is why near-infrared light, corresponding to frequencies of hundreds of terahertz, is interesting for communications. Optical communications allow rates of many terabits per second.

Furthermore, a high-frequency band allows the allocation of plenty of channels, yielding no need for spectrum allocation constraints. This benefit adds to the capacity of sending big amounts of high-resolution data in a reduced time, and it is the basis of current optical communications.

An optical fiber for long-distance communication is a glass (silica) wire, a bit thinner than a human hair, consisting of a highly transparent core and a cladding cover with a different amount of impurities. The **refractive index** of the core is slightly higher than that of the cladding, so light travels along the fiber confined inside by total internal reflection. The favored wavelength for fiber-optical communication is 1550 nm, which corresponds to the region in which attenuation is minimal, thanks to the low values in the **absorption spectrum** and scattering of silica.

Since the first transatlantic fiber-optic cable was installed in 1988, the number of optical fiber cables all over the world is constantly increasing. Nowadays a single fiber can hold in a second millions of phone conversations and tens of thousands of TV channels.

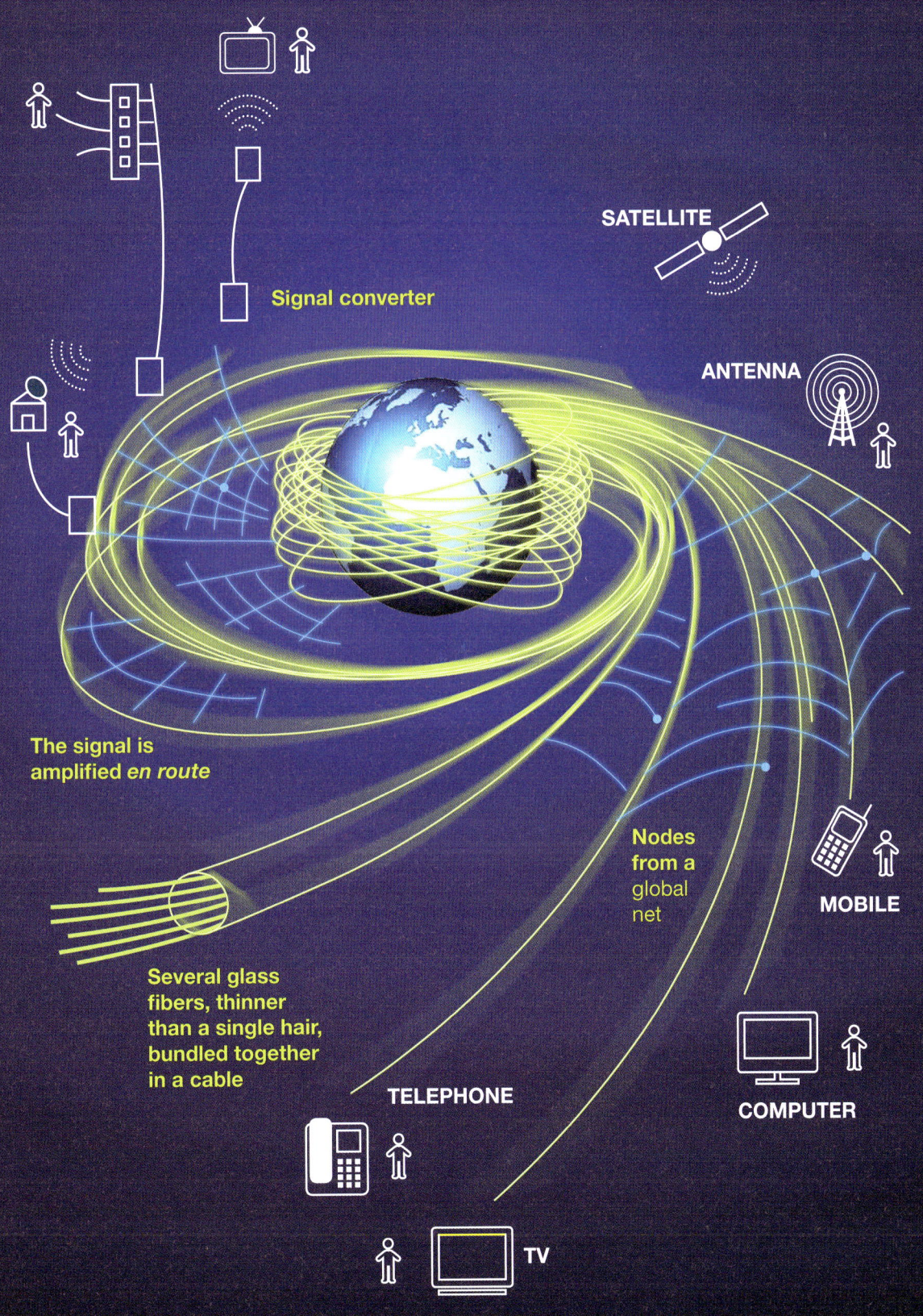

Signal converter

SATELLITE

ANTENNA

The signal is
amplified *en route*

Nodes
from a
global
net

MOBILE

Several glass
fibers, thinner
than a single hair,
bundled together
in a cable

COMPUTER

TELEPHONE

TV

Scientific and technical advisors

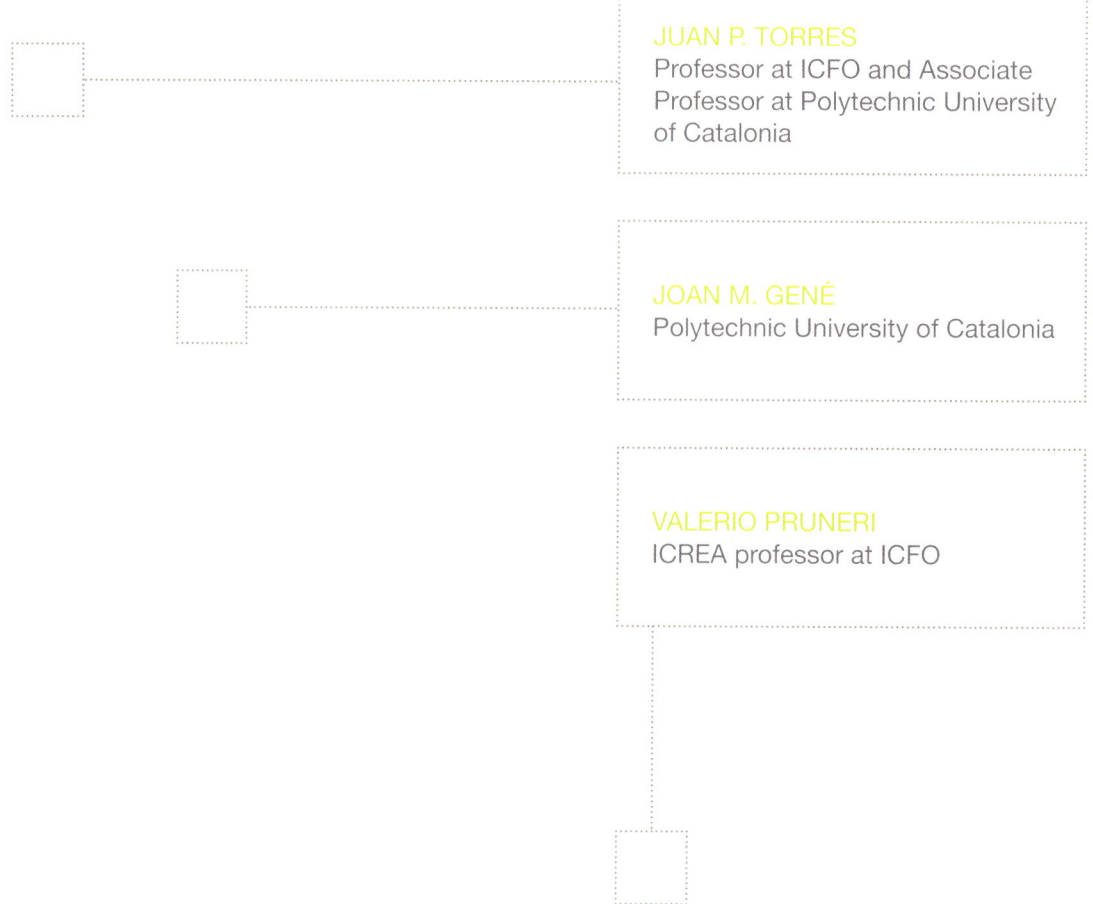

JUAN P. TORRES
Professor at ICFO and Associate
Professor at Polytechnic University
of Catalonia

JOAN M. GENÉ
Polytechnic University of Catalonia

VALERIO PRUNERI
ICREA professor at ICFO

Glossary

ABSORPTION SPECTRUM A pattern of dark lines observed in the spectrum of light when it passes through a particular absorbing medium, corresponding to the missing frequencies that have been absorbed. Such a pattern is unique and can be used to identify the medium.

BANDWIDTH The range of frequencies that a transmission physical channel supports. It is measured in Hz.

FREQUENCY In physics, the number of waves that pass a fixed point in a unit of time.

REFRACTIVE INDEX Measure of the bending of a ray of light when passing from one medium into another.

Definitions adapted from
Oxford American English Dictionary
Encyclopaedia Britannica

Relevant reading

Dourish, P. Bell, G. (2011) Divining a Digital Future. MIT Press, Cambridge (MA)

Hecht, J. (1999) *City of Light. The Story of Fiber Optics.* Oxford University Press, London

The 2009 Nobel Prize in Physics – Popular Information. Nobelprize.org. Nobel Media AB 2014. Web. 8 Jul. 2014. http://www.nobelprize.org/nobel_prizes/physics/laureates/2009/popular.html

The Nobel Prize in Physics 2009 – Advanced Information. Nobelprize.org. Nobel Media AB 2014. Web. 8 Jul. 2014 http://www.nobelprize.org/nobel_prizes/physics/laureates/2009/advanced.html

Mims, F. M. (1981) Alexander Graham Bell and the Photophone. *Optics News* **6(8)**

NASA Communications Relay Demonstration http://esc.gsfc.nasa.gov/267/LCRD.html

Weiser, M. (1991) The computer for the twenty-first century. *Scientific American* **265(3)**: 94–104

Photo credits

p 76   Cable-laying ship © Ocean Networks Canada 2015

p 77   Communications poster, adapted with permission of The Royal Swedish Academy of Sciences

PRIVACY

The key is quantum

PRIVACY

PRIVACY

Beyond approximately 10 years into the future, the general feeling among ECRYPT partners is that recommendations [on digital information security] made today should be assigned a rather small confidence level.
2008 ECRYPT European Network of Excellence in Cryptology final report.

An old saying, very popular among cryptographers, states that two people can share a secret only if one of them is dead; there is nothing that makes a piece of information more compelling than keeping it secret.

The need for private communications might seem today as momentous as ever, with an increasing number of our relevant activities digitalized and uploaded to the cloud. But keeping secrets undercover has always been of paramount importance. The wealth and fate of whole projects and enterprises have often depended on the competence of skillful agents in defending confidential information, even with their lives.

Cryptography is indeed as ancient as the civilized world. Its history is frequently pictured as an endless fight between the information hiders and the information seekers: the code-makers and the code-breakers. Each battle, eventually won by the increasing ingenuity of code-breakers, which forces in turn the springing up of yet craftier code-makers, stimulates the rise and expansion of several scientific disciplines. Some of the finest minds of all ages have been fascinated by the challenge of deciphering codes: Al-Kindi in the Middle Ages, Charles Babbage in the nineteenth century, and Alan Turing in the twentieth century, were prominent code-breakers and made fundamental advances in cryptanalysis.

A priceless by-product of all these advances was a deep progress in other areas such as linguistics, statistical analysis, and computer science.

Right now, at the forefront of code-making, stand Ron Rivest, Adi Shamir, and Leonard Adleman. They are the inventors of the RSA cryptosystem that maintains security on the Internet. The system, based on the practical difficulty of factoring the product of two big prime numbers, has brought higher mathematics into the cipher arena, where we now find elements of number theory, modular arithmetic, or elliptic curves. But this kind of mathematics may not yield the final solution to a secure transmission of information, if there is any. Indeed, the security of RSA is based on the unproved assumption that no algorithm is capable of solving a certain mathematical problem in a timely fashion. RSA is thus threatened by the discovery of such an algorithm or by the sprouting of much faster computers. Quantum computers – although not yet functional – may in the near future become the ideal allies of code-breakers as they could decode efficiently all RSA-encoded messages, especially the ones that are being transmitted as we speak.

But quantum physics is also here to serve the code-makers. And it might be their definitive strike. The security of quantum cryptography is not based on unproven assumptions: it is guaranteed by the laws of physics. Physics is the last field fertilized by the advance of cryptography. All we need now is to find a way of sending information under the loyal protection of the strange properties of photons, which, much in the style of ancient couriers, will sacrifice their physical integrity if threatened by a spy attack.

Problem:
how to send a secret message.

A simple method to transmit secret information is sending a written message hidden under a physical veil.

Very clever keys have been invented over the centuries to conceal the meaning in a jumble of unintelligible words.

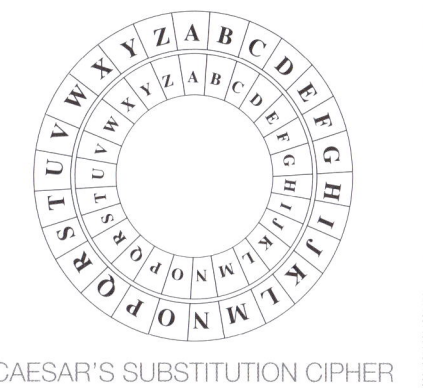

CAESAR'S SUBSTITUTION CIPHER

The problem is
how to send the secret deciphering key.

After a millenary fight against spies, cryptographers might stand a chance of winning for good if they return to principles and find a way to hide information in (quantum) physical boxes.

ALICE

BOB

EVE, THE SPY

PRIVACY

The key is quantum

Quantum cryptography is based on the one-time-pad: two parties, Alice and Bob, share a private key K, i.e. a random sequence of 0s and 1s. Alice then encodes her message, *or plaintext* M (a sequence of 0s and 1s) in a *ciphertext* C which is the **binary sum** (0+0=0; 1+0=1; 1+1=0) of M and K: M+K=C. C is sent to Bob, who uses the key again to recover the original message M=C+K (since K+K=0).

👍 The one-time-pad has two great assets: (a) the *ciphertext* C can be safely sent through a public channel, and (b) it is provably 100% secure.

👎 It has two serious caveats: (a) the key has to be as long as the message, and if the key is used more than once most information on the key and messages is exposed, and (b) the prerequisite of sharing a (long) random secret string is impractical.

Quantum key distribution provides Alice and Bob with a secret random key. Its security is rooted in a tenet of quantum mechanics: measurement inevitably disturbs the system. The slightest attempt of a spy to read the information exchanged between the two parties is immediately detected by them. In that case, they throw away the key and stand by until the eavesdropper stops listening. Otherwise they proceed using the one-time-pad rendering an unconditionally secure protocol.

Quantum information carriers par excellence are **photons**: **laser** sources provide photons with well-defined properties which speedily travel in optical fibers or in open space, interacting very weakly with their environment. Information can be encoded in the **polarization** state of photons.

Classical cryptography

is based on the assumption that the spies are not clever enough to break the key nor cunning enough to steal it.

Quantum cryptography

is based on fundamental laws of Nature.

Unpredictable quantum events generate sequences with no pattern to be discovered, and no room for astute spies either, for a quantum system can tell when it's being observed.

Scientific and technical advisors

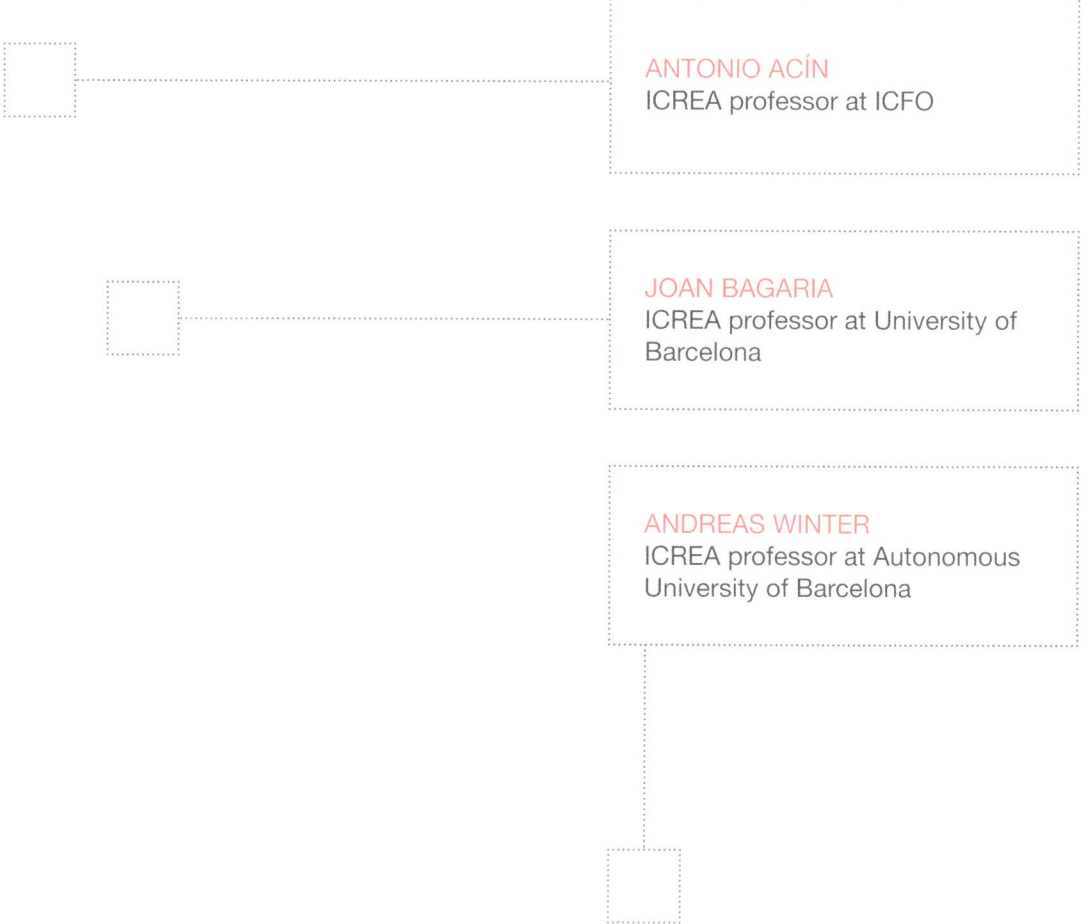

ANTONIO ACÍN
ICREA professor at ICFO

JOAN BAGARIA
ICREA professor at University of
Barcelona

ANDREAS WINTER
ICREA professor at Autonomous
University of Barcelona

Glossary

....

BINARY SUM   The binary sum + is done bit-by-bit returning 0 if both bits are the same or 1 if they are different.

LASER   Acronym for Light Amplification by Stimulated Emission of Radiation. A laser produces a highly coherent, highly directional, and nearly monochromatic beam of light, by stimulating atoms or molecules to emit light at particular wavelengths and amplifies that light, typically producing a very narrow beam of radiation.

PHOTON   A photon is a light quantum, an energy packet of electromagnetic radiation.

POLARIZATION   In a beam of electromagnetic radiation, the polarization direction is the direction of the oscillation of the electric field of light. A photon has a polarization state that can be written as a linear combination of two distinct polarization states, e.g. horizontal and vertical polarization, in which information can be encoded.

....

Definitions adapted from
Oxford American English Dictionary
Encyclopaedia Britannica

Relevant reading

Brassard, G. (2006) Brief history of quantum cryptography: a personal perspective. *Proceedings of IEEE Information Theory Workshop on Theory and Practice in Information Theoretic Security*, Awaji Island, Japan. Available online at http://www.iro.umontreal.ca/~brassard

Cox, B., Forshaw, J. (2011) *Everything That Can Happen Does Happen.* Penguin Books, London

Gardner, M. (1972) *Codes, Ciphers and Secret Writing*. Simon & Schuster, New York

Kahn, D. (1996) *The Code Breakers – The Comprehensive Story of Secret Communication from Ancient Times to the Internet*. Scribner, New York

Singh, S. (1999) *The Code Book.* Doubleday, New York

Photo credits

p 84   Artist conception of laser secure communications © ICFO-Digivision

p 85   Enigma machine H. Photograph reproduced with permission of the Crypto Museum www.cryptomuseum.com

RIDDLES

Quantum computation

RIDDLES

A problem frowns, a riddle lifts an eyebrow,
a paradox smiles:
– X believes she is a hypochondriac.
Is X a hypochondriac?

Define hypochondria as the unfounded belief of having a disease. Assume hypochondria is a disease. Then if X above is a hypochondriac, then X is not a hypochondriac. But if X is not a hypochondriac, then X is a hypochondriac.

There is something liberating in discovering flaws in rigid systems, as long as they don't threaten your own security. A paradox is a flaw on the rigid system of reasoning. Any formalization of reasoning can be pictured as a path to the top of a mountain. To answer a question, one has to climb to the right solution stepping on single rocks, which are intermediate states of certainty. A paradox is fun because it means bouncing forever between two rocks: it shows that the Mountain of Truth can be tricked. Although this is serious enough to deeply shake the foundations of mathematics, causing one of the greatest philosophical revolutions ever, for the layperson it is a guarantee that playing games on thought will never be over.

Paradoxes are not the only examples of exceptional climbing up the Mountain of Truth. There are questions with infinitely long paths of rocks, or with no path at all. There are questions with paths which, being finite, are so lengthy that it is not possible to reach the peak in trillions of years. If a given question is proved to have no answer, then one is allowed to devote efforts to more efficient tasks, but if a question has an answer and the problem is simply that all the time in a lifetime is not enough to reach the peak, then Mountain-of-Truth climbing deserves a revision. Classical Mountain-of-Truth climbing is characterized by the fact that one always steps on a single true rock at a time. To jump from one rock to another, one just has to give basic true/false answers. This finds a correspondence in classical physics, where things can be in only one place at a time. And it also finds a correspondence in logic, where sentences cannot be true and false at the same time. This correspondence is the essence of the ability of classical machines to carry out classical computations by following classical algorithms.

How would algorithms change if supported by non-classical physics, say a set of laws that allow things to be in different places at a time? This is one of the questions that physicists speculated about when confronted with the task of understanding the particular revolution that shook the foundations of their field, causing yet another major philosophical revolution: quantum physics. This question goes right to the origins of quantum computation.

A quantum climber may step on several rocks at a time, and may take a step in different directions at the same time. This means that he can quickly explore the truth of a vast number of paths. This and other puzzling – and yet real as a rock – features of the quantum climber can be used to provide in human timescales a solution to problems that would require cosmological timescales to any classical computer.[1]

Now, the challenge is to build a quantum computer. It is truly a complex feat to make real physical objects take simultaneous steps, and even more demanding to keep them there, in a so-called superposition state, during enough time to finish the rest of the calculation. It seems photons have the right aptitudes to face the Mountain of Truth if the rocks are made out of the right atoms. Which path the climbers took is a nice riddle to think about.

---

[1] For instance, while no classical algorithm is known to solve in an efficient time frame the problem of finding the prime factors of a sufficiently large number – which is the basis of Internet security – Peter Shor's quantum algorithm could do it, if programmed in a sufficiently big quantum computer.

Classical algorithms solve problems routinely proceeding step by step:

Start!

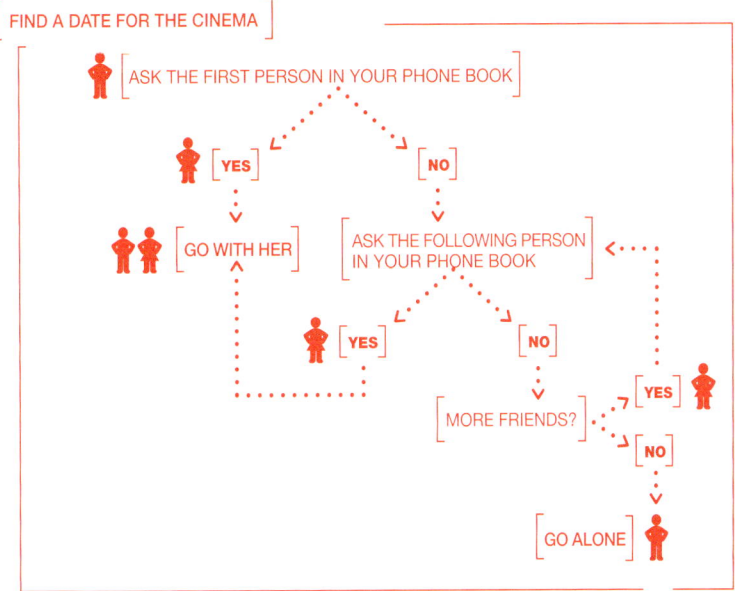

FIND A DATE FOR THE CINEMA

ASK THE FIRST PERSON IN YOUR PHONE BOOK

YES

NO

GO WITH HER

ASK THE FOLLOWING PERSON IN YOUR PHONE BOOK

YES

NO

MORE FRIENDS?

YES

NO

GO ALONE

But some hard mathematical problems would take several lifetimes of the Universe to finish, if calculated step by step:

FIND THE PRIME FACTORS OF $n = (2^{57885161} - 1)(2^{37156667} - 1)$

Quantum algorithms operate in a fundamentally different way, using the resources of several parallel "universes" simultaneously, so computation time significantly decreases.

The challenge is how to manipulate matter and light to read the correct solution in this Universe.

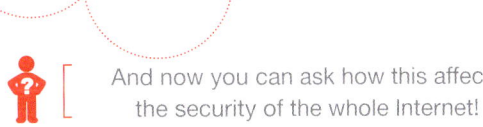

And now you can ask how this affects the security of the whole Internet!

RIDDLES

Quantum computation

The information managed by a quantum computer comes in quantum bits, or qubits. A qubit can be encoded in a two-dimensional quantum system, for instance, in the **polarization** state of a single photon. The logical states 0 and 1 can correspond to a single photon being left- and right-circularly polarized, $|0\rangle$ and $|1\rangle$, respectively. A general qubit state can be expressed as a **superposition** of the states $|0\rangle$ and $|1\rangle$. General states of quantum information are superpositions of strings of qubits.

In an optical quantum computation, qubits are registered and retrieved from the memory through quantum light–matter interactions. The role of the quantum memory is to store the superpositions of qubits in **quantum coherent** atomic states, allowing the synchronized and parallel processing of many qubits. Storage time is therefore limited by **decoherence**.

A basic scheme consists of a photon in an arbitrary polarization state and an atom in the ground state. The energy of the photon matches an electronic transition of the atom. When the photon is absorbed by the atom, polarization is mapped to an atomic superposition. Then a short control pulse transfers such a superposition to a long-lived ground state, which is stored as long as the superposition is maintained (i.e, decoherence is avoided). Retrieval is obtained using another short control pulse that reverses the process. For the interaction to be efficient, either an ensemble of atoms is used or the atom is placed on a **high-finesse cavity**, which confines the photon.

Several protocols and several atomic systems have been used to implement quantum memories, including laser-cooled atomic ensembles, atomic vapors, single atoms in high-finesse cavities, and atomic ensembles in **rare-earth** ion doped crystals.

Quantum memories for light also have other applications: in precision measurements, quantum information repeaters, single-photon detection, or tests of the foundations of quantum mechanics.

$$|\Psi\rangle = \sqrt{p}\,|0\rangle + e^{i\vartheta}\sqrt{1-p}\,|1\rangle$$

Dealing with quantum information raises the challenge of storing during an appreciable amount of time something as ephemeral as a superposition of states.

A quantum particle

may exist in various states simultaneously,
eluding for a certain amount of time the decision of which of these states to choose…

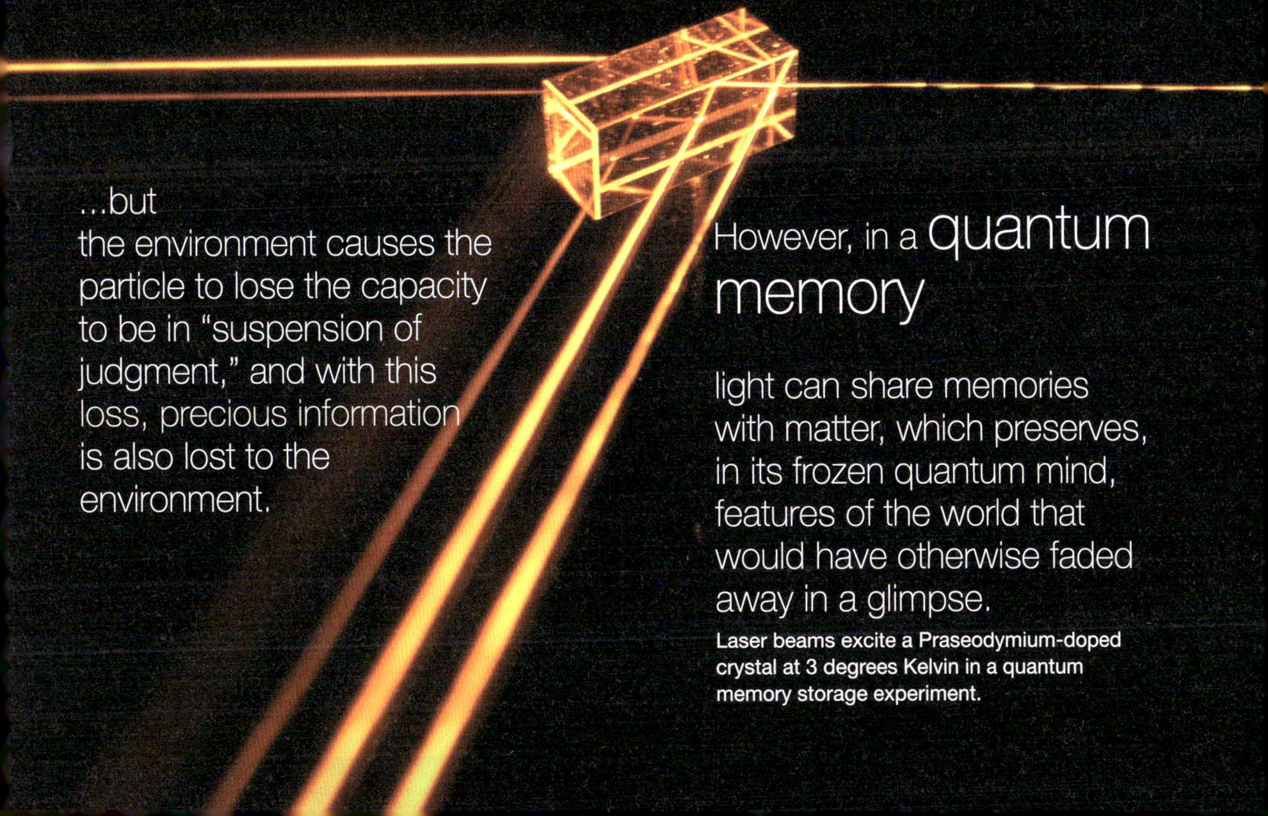

…but
the environment causes the particle to lose the capacity to be in "suspension of judgment," and with this loss, precious information is also lost to the environment.

However, in a quantum memory

light can share memories with matter, which preserves, in its frozen quantum mind, features of the world that would have otherwise faded away in a glimpse.

Laser beams excite a Praseodymium-doped crystal at 3 degrees Kelvin in a quantum memory storage experiment.

Scientific and technical advisors

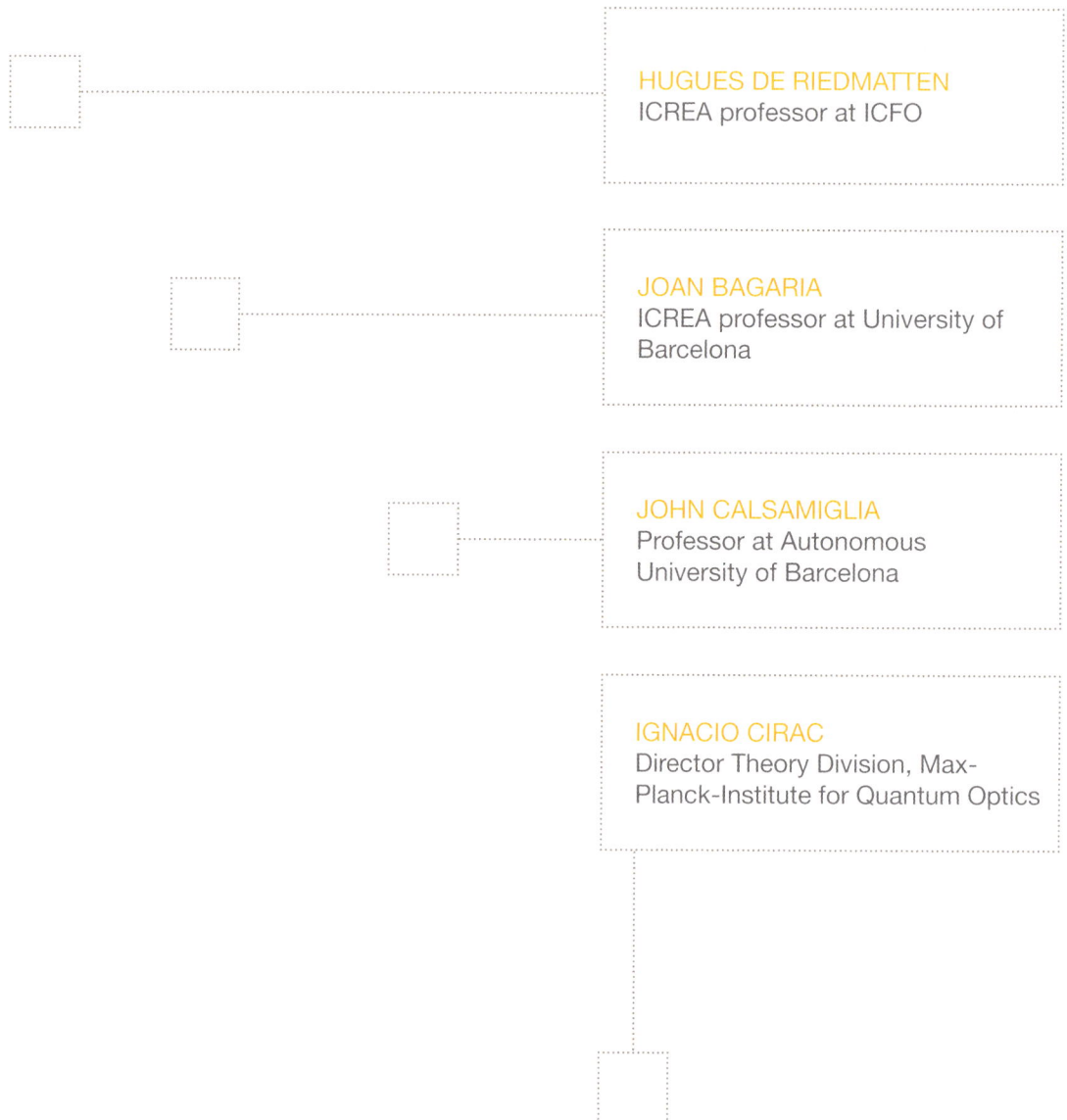

HUGUES DE RIEDMATTEN
ICREA professor at ICFO

JOAN BAGARIA
ICREA professor at University of Barcelona

JOHN CALSAMIGLIA
Professor at Autonomous University of Barcelona

IGNACIO CIRAC
Director Theory Division, Max-Planck-Institute for Quantum Optics

Glossary

COHERENCE In quantum mechanics, objects have wave-like properties. Two waves of the same frequency are said to be coherent when their relative phases over time are constant, rendering stable interference patterns between them visible. A quantum particle is said to be in a coherent superposition of two quantum states when the different terms in the superposition have a fixed relative phase, rendering quantum effects like "being in two places at the same time" visible.

DECOHERENCE Any process which effectively scrambles the phase difference between terms in a quantum superposition. Superpositions of states become classical probabilistic mixtures. Features like interference or entanglement, processes like quantum teleportation, or applications like quantum computation are hampered by the presence of decoherence.

HIGH-FINESSE CAVITIES A trap for photons consisting in two slightly convex opposing mirrors.

POLARIZATION In a beam of electromagnetic radiation, the polarization direction is the direction of the oscillation of the electric field of light. A photon has a polarization state that can be written as a linear combination of two distinct polarization states, e.g. horizontal and vertical polarization, in which information can be encoded.

RARE-EARTH ELEMENT Any chemical element of the first extended row below the main body of the periodic table (cerium through lutetium) plus scandium and yttrium.

SUPERPOSITION The principle of superposition: A quantum system that can be in two different states $|0\rangle$ and $|1\rangle$, can also be in a superposition of those $|\psi\rangle = a|0\rangle + be^{i\phi}|1\rangle$ where a is the probability amplitude of finding the system in $|0\rangle$, b is the probability amplitude of finding it in $|1\rangle$, and ϕ is the relative phase.

Definitions adapted from
Oxford American English Dictionary
Encyclopaedia Britannica

Relevant reading

Aaronson, S. (2013) *Quantum Computation Since Democritus*. Cambridge University Press, Cambridge

Bussières, F., Sangouard, N., Afzelius, M. *et al.* (2013) Prospective applications of optical quantum memories. *Journal of Modern Optics* **60(18)**: 1519–1537

Cox, B., Forshaw, J. (2011) *Everything That Can Happen Does Happen*. Penguin Books, London

Lvovsky, A. I., Sanders, B. C., Tittel, W. (2009) Optical quantum memory. *Nature Photonics* **3**: 706–714

Scarani, V. (2006) *Quantum Physics, a First Encounter: Interference, Entanglement, and Reality*. Oxford University Press, Oxford

Photo credits

p 92   Cryogenically cooled solid state quantum memory © 2012 ICFO, picture by Kutlu Kutluer

p 93   Laser beams on Praseodymium © 2010 Matthew Sellars, University of Canberra

NEW MATERIALS

Physics lessons from the miniatures

NEW**MATERIALS**

The term "material world" refers in some contexts to the limitations of our will, the wall on which ideas crash to become just reality. And yet materials are – with all their imperfections – one of the ultimate fabrics for testing our thoughts. Fortunately, between the world of pure ideas and the actual material reality, there exists an exciting third region: the region of possibilities. This region is occupied by materials not forbidden by Nature that, however, do not actually occur (or occur very rarely). It contains everything we can soundly think of, but that is too odd as to have been inspected by Nature. A family of such entities is formed by what is termed "metamaterials."

Optical metamaterials are man-made materials whose peculiar properties emerge from an artificial inner structure, carefully crafted at the nanoscale. It is possible (yet not easy) to produce materials whose fine structure tricks visible light to bend in unconventional (yet not impossible) ways. These severe alterations of the refraction properties of a material might produce supermagnifying lenses – with enough power to see directly a DNA chain – or even guide the trajectory of light around an object and render it invisible: light would illuminate the metamaterial covering the object and travel along it to emerge to our eyes in the same direction and carrying the same information it had prior to illumination. No reflection, no absorption: a true invisibility cloak.[1]

Such response of light to structure is possible because light is an electromagnetic wave. Indeed light, as described by Maxwell's laws, consists of an electric field and a magnetic field whose spatiotemporal variations are related to each other, to their sources, and to the characteristics of the medium through which it propagates. If the "tricks" introduced in the structure of the material are smaller than the wavelength of light, then, although microscopically the passage of light through the material is extremely intricate, macroscopically it can be described in terms of the new characteristics of the effective medium, that turn out to be extraordinary. Invisibility cloaks and supermagnifying lenses, although amazing, are still in their infancy. However, some less demanding artificial guiding of light, with still-important short-term applications, can be found for example in photovoltaics, where such guiding might help for an enhanced collection of light by a solar cell.

The study and manipulation of the behavior of light at the nanoscale constitutes a broad field known as nanophotonics, and includes research areas beyond the development of optical metamaterials. Very small metallic particles, of size comparable to the wavelength of visible light, can collect or direct visible light, just as standard antennas do for radiowaves. Depending on their size and form, these nanoparticles can also focus the electric field of light and act as mini-tweezers, or concentrate heat and act as mini-ovens, or simply resonate with selected frequencies of the incoming light to produce the exceptionally brilliant colors of the rose windows in medieval churches.

We devote the two illustrated pages of this chapter to such senior material, the first man-made material that showed extraordinary properties based on its nanostructure, long before the birth of nanotechnology rescued some more new materials from the region of possibilities.

---

[1] Note that researchers have already demonstrated the principle, over a certain range of angles, using conventional optics.

Maxwell equations

1. $\nabla \cdot E = \dfrac{\rho}{\varepsilon_0}$

2. $\nabla \times E = -\dfrac{\partial B}{\partial t}$

3. $\nabla \times B = \mu_0 J + \mu_0 \varepsilon_0 \dfrac{\partial E}{\partial t}$

4. $\nabla \cdot B = 0$

The ways in which light is reflected

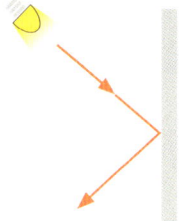

and refracted

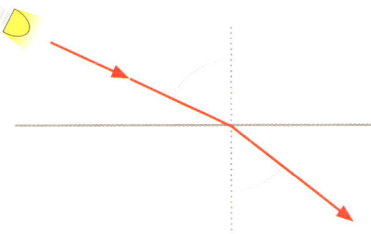

in a material...

...are determined by the **ε**lectric and **μ**agnetic constants of the material.

Careful engineering at the nanoscale of **ε** and **μ** turns a material into a metamaterial.

In an optical metamaterial, light bends in unconventional directions to get unconventional properties.

There are other ways in which a nanostructure can affect the behavior of light.

NEW MATERIALS

Physics lessons from the miniatures

Optical metamaterials are built up from mini-building blocks, much smaller than the **wavelength** of visible light (approximately 500 nm) but still several orders of magnitude bigger than atoms. They are engineered to equip a nanostructure with the capacity of affecting the propagation of light.

The propagation of light through a medium, as explicitly shown by Maxwell equations, is determined by the medium's **permittivity**, ε, and **permeability**, μ. These parameters are positive in any standard material, producing positive **refraction** angles, a positive bending of light when entering the medium. A specially designed metamaterial can in principle produce negative refraction angles, with prospects of amazing applications, such as super-lenses and invisibility cloaks. Such negative **refractive index** materials were among the first proposals of metamaterials; however, nowadays the meaning of the term has broadened to generally include any nanostructure with the ability to concentrate the intensity of light or twist the trajectory of light rays.

Metallic nanoparticles, which we separate here from metamaterials, involve free electrons that resonate with the electric field of light, resulting in a concentration of light in very small regions. The size, the form, and the specific metal of the nanoparticle determine its response to light.

Once the light is concentrated in such small regions, it can be applied in super-resolution microscopy, to probe nanoscopic details of objects; or in sensing, to detect substances whose presence near the nanoparticle would shift its plasmonic resonance and hence the wavelength of light; or in medical sciences, to non-invasively destroy tissue by a concentration of heat; or in photovoltaics, to enhance the collection efficiency of solar cells.

Optical metamaterials and nanoparticles both fall under the denomination of new materials because the obtention of the involved nanopieces is only possible thanks to spectacular advances in nanofabrication that occurred during the past few decades, including several nanolithography techniques (e-beam lithography, **laser** writing) and colloidal chemistry.

Metallic nanoparticles!

TRANSMISSIONS ELECTRON MICROGRAPHS OF:

PHOTOGRAPHS OF COLLOIDAL DISPERSIONS OF:

Gold–silver nanospheres (citrate reduction).

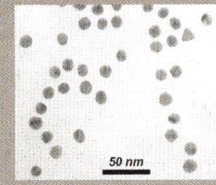 

Increasing Au concentration.

Gold nanorods (seeded growth).

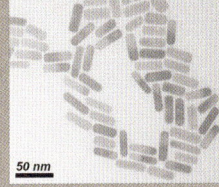 

Increasing aspect ratio.

Silver nanoprisms (DMF reduction).

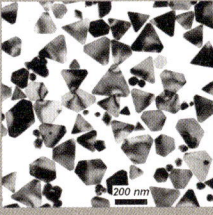 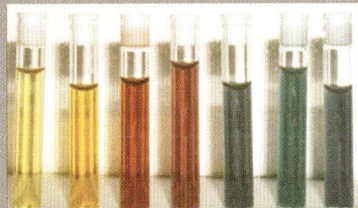

Increasing lateral size.

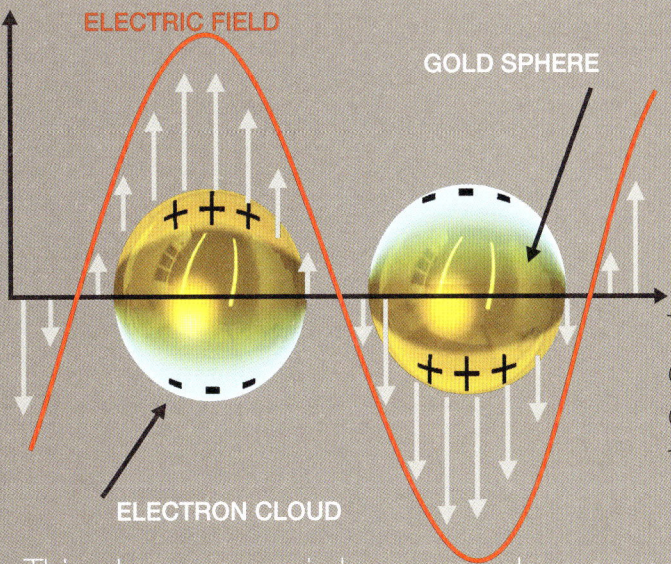

When the size of a metallic particle is much smaller than the wavelength of light,

its electron cloud oscillates with the electric field of light for a well-defined frequency band.

ELECTRIC FIELD

GOLD SPHERE

ELECTRON CLOUD

TIME

This oscillation makes the particle very bright.

The resultant color depends on the size of the particle.

This phenomenon is known as plasmon resonance.

Scientific and technical advisors

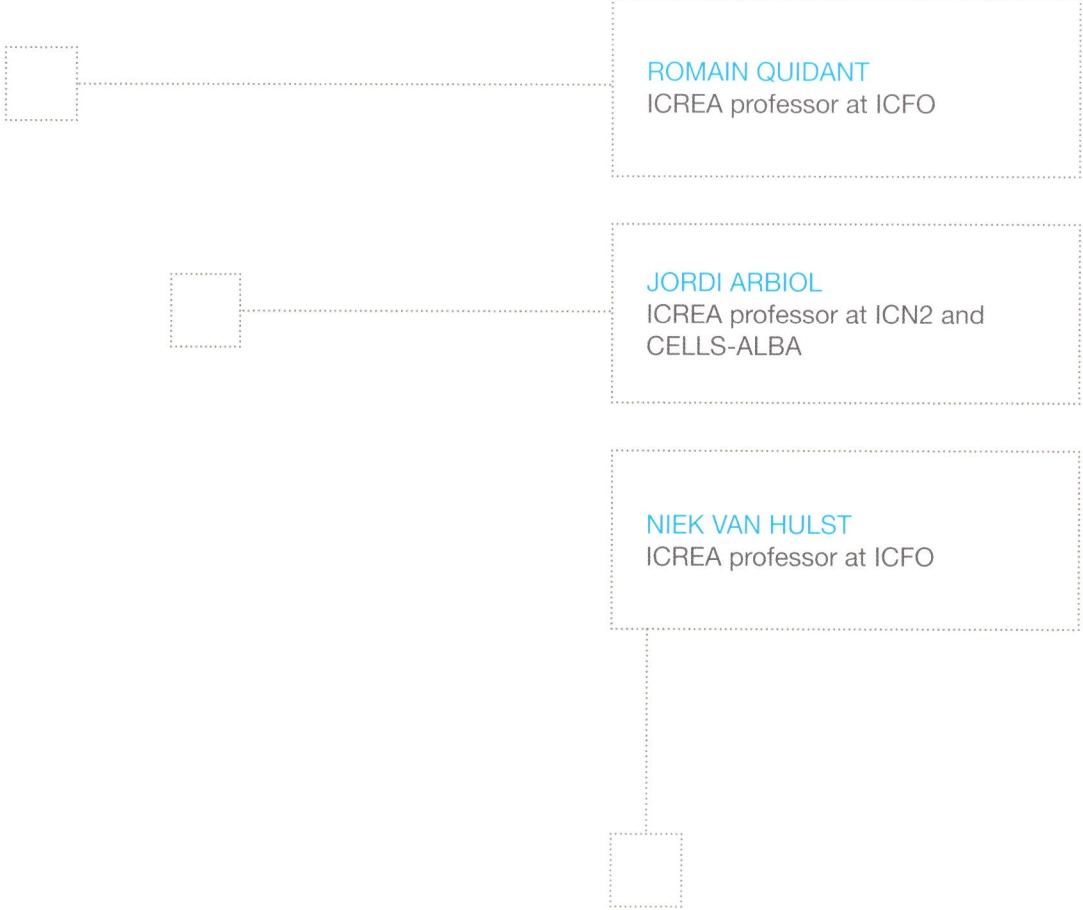

ROMAIN QUIDANT
ICREA professor at ICFO

JORDI ARBIOL
ICREA professor at ICN2 and
CELLS-ALBA

NIEK VAN HULST
ICREA professor at ICFO

Glossary

....

LASER Acronym for Light Amplification by Stimulated Emission of Radiation. A laser produces a highly coherent, highly directional, and nearly monochromatic beam of light, by stimulating atoms or molecules to emit light at particular wavelengths and amplifies that light, typically producing a very narrow beam of radiation.

PERMEABILITY Relative increase or decrease in the resultant magnetic field inside a material as a response to an applied magnetic field.

PERMITTIVITY Constant of proportionality that relates the electric field to the electric displacement in a material.

REFRACTION The bending of oblique incident rays as they pass from a medium having one refractive index into a medium with a different refractive index.

REFRACTIVE INDEX Measure of the bending of a ray of light when passing from one medium into another.

WAVELENGTH Electromagnetic energy is transmitted in the form of a sinusoidal wave. The wavelength is the physical distance covered by one cycle of this wave.

....

Definitions adapted from
Oxford American English Dictionary
Encyclopaedia Britannica

Relevant reading

Leonhardt, U. (2007) Optical metamaterials – invisibility cup. *Nature Photonics* **1**: 207–208

Pendry, J. B., Smith, D. R. (2006) The quest for the super-lens. *Scientific American* **295**: 60–67

Wagner, F. E., Haslbeck, S., Stievano, L., *et al*. (2000) Before striking gold in gold-ruby glass. *Nature* **407**: 691–692

Drexler, K. E. (1992) *Nanosystems: Molecular Machinery, Manufacturing, and Computation*. Wiley, New York

Stephenson, N. (1995) *The Diamond Age*. Bantam Books, New York

Atwater, H. A. (2007) The promise of plasmonics. *Scientific American* **17**: 56–63

Photo credits

p 100 Rose window at Strasbourg Cathedral, courtesy of Clostridium

p 101 Image of solutions, courtesy of Luis Liz Marzán, Chemical Colloidal Group, University of Vigo

CATCH THAT ENERGY!

A pile of connected events in the flow of photovoltaics

CATCHTHAT
ENERGY!

As we walk through city streets, we walk through time, encountering the city-building legacy of each past generation.
Cliff Ellis, *History of Cities and City Planning*

In the 1980s, American city planner Kevin Lynch identified in the book *Good City Form* five basic dimensions by which to evaluate a city: *vitality, sense, fit, access*, and *control*. It is our impression that if city performance is ever to be measurable, something similar to Lynchean *vitality* is very likely to appear among its defining parameters.

Vitality is defined in this context as the ability to fulfill the biological needs of citizens, while providing a safe and healthy environment. In contrast, cities are nowadays responsible for approximately three-quarters of the world's energy consumption, and face the consequent problems related to overpopulation, exhausted energy resources, and pollution. Abandoning the city model and hence scattering in small villages is not an option. People and ideas flowing in cities are at the root of creativeness, innovation, and transformation that spread through all of the civilized world. City planners (and also current residents, through a rational use of energy resources) are the ones endowed with the task of assuring the vitality of citizens for generations to come.

Securing enough resources, especially clean and renewable energy, is one of the challenges for city planners. The job is not trivial, since most cities are already built. Founded by crossroads, rivers, or natural harbours to guarantee their energy supply, cities have grown along with the availability of new sources of energy which contribute to their nurturing. New energy is transported into the city from wind farms, solar plants, nuclear plants, oil wells, or gas

reservoirs, installed not always around the corner, with a growing environmental, economic, and energetic cost.

Out of all of the above sources of energy, there is one that could be directly integrated into the city with no need for transportation: solar power. The Sun provides the Earth with as much energy every hour as the worldwide energy consumption over a year (data from 2002). The efficiency in energy conversion is mainly limited by the current photovoltaic technology, but there are other factors to take into account, like latitude, season of the year, or orientation. All in all, optimizing the solar contribution to the energy of the city amounts to maximizing the efficiency of solar cells and/or the size of the surface area of solar cells.

A city planner would look at the city map: buildings, networks (streets), and open places (parks). She would, in the majority of cases, discard placement of the solar cells in parks, and on the streets. Buildings would appear as one of the best options for integrating photovoltaics in the city. However, the area offered by the roofs might prove insufficient. By a process of elimination, we are left with the possibility of covering buildings' façades, perhaps even the windows! This proposal needs a more versatile technology that produces large sheets of thin, light, flexible, and, if necessary, transparent photovoltaic cells. On top of that, the process should be economically, energetically, and environmentally viable.

We just stated the objectives of transparent integrated photovoltaics: elucidating a set of new photonic materials, structures, and fabrication methods to be incorporated into the future urban energy systems. With luck, this will be a small part of our *city-building legacy* for the next generation.

On an average day on Earth,
energy consumption peaks
when solar irradiance peaks.

(That's only natural: most people are
busier during the central hours of the day.)

So it's only natural that we
want to use solar energy.

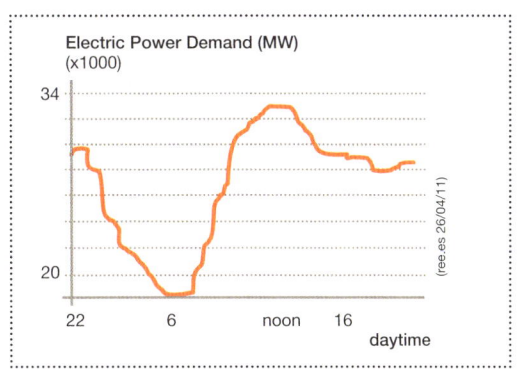

Electric Power Demand (MW)
(x1000)

34

20

22   6   noon   16

daytime

(ree.es 26/04/11)

And the challenge
is how to get the
most of it…

…with a whole set
of new photovoltaic
technologies.

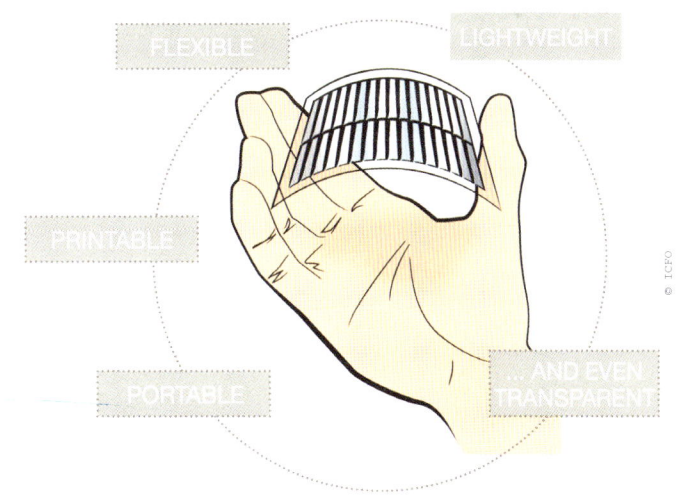

FLEXIBLE   LIGHTWEIGHT

PRINTABLE

PORTABLE   … AND EVEN
                  TRANSPARENT

© ICFO

SOLAR CELLS

CATCH THAT ENERGY!

A pile of connected events in the flow of photovoltaics

Solar cells convert sunlight into electricity by virtue of the photovoltaic principle, by which electrons in the valence band of a material take the energy of the photons to jump to the conduction band. The basic scheme of a solar cell is a pile of different materials sandwiched between two electric contacts. The core of the cell consists of two layers of a semiconductor material (typically **silicon**). One of these layers (n-type silicon) has an excess of electrons and the other (p-type silicon) has an excess of holes (the absence of electrons). When this junction is illuminated and short-circuited a current in the reverse direction flows (see figure below).

Solar cells have large surface areas to maximize the exposure to the Sun. Typically, they absorb in the **visible** and near-**infrared** regions of the solar spectrum, the region of maximum solar irradiance. Most of the commercialized solar cells are based on crystalline silicon, with a current efficiency of around 15% or 16%.

New photonic materials offer properties that can incorporate interesting functionalities to photovoltaics. Transparent, thin, flexible solar cells might be attached to any surface, for instance to buildings' façades – including windows – to maximize the surface area. An example of such technology includes nanometric layers of blends of solution-processed organic or inorganic materials as photovoltaic materials; nanostructured materials designed to enhance absorption, both by extending the absorption band to the near-infrared and the near-**ultraviolet**, and by keeping light bouncing inside; and even special patterns to guide light, compensating the low irradiation on vertical structures and augmenting significantly the amount of absorbed light.

Although the efficiency of these technologies is currently lower than standard photovoltaics, in the future they may turn out to be very cost-effective for large scale deployment.

A pile of alternating pieces of zinc and silver in a weak acid solution produced the first chemical-based electric current.

PHOTOVOLTAIC EFFECT

The energy of photons makes electrons in the valence band jump to the conduction band. In other words, photons create electricity!

In 1800 Alessandro **Volta**
invented the battery.

In 1801 Napoleon created
the **Volta Prize** to honor such a contribution
of science to society.

In 1880 Alexander Graham Bell was awarded
the **Volta Prize** for the invention of the
telephone.

With the prize money, Bell founded
Volta Labs, later known as Bell Labs.

In 1954 Bell Labs presented the first modern
solar cell based on the **photovoltaic** effect.

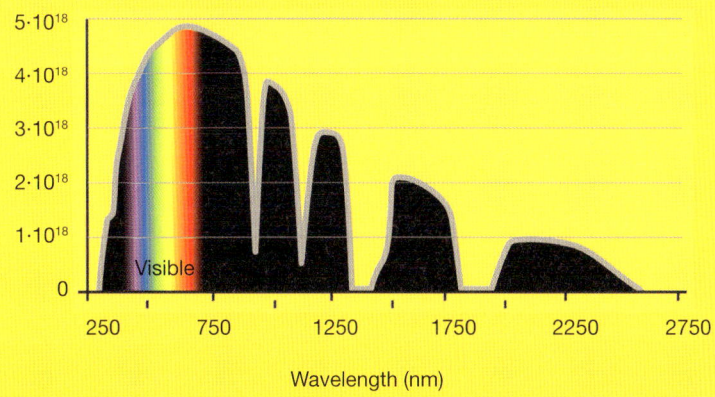

PHOTON FLUX ON EARTH: NUMBER OF PHOTONS/m²·nm·s

Visible

Wavelength (nm)

Scientific and technical advisors

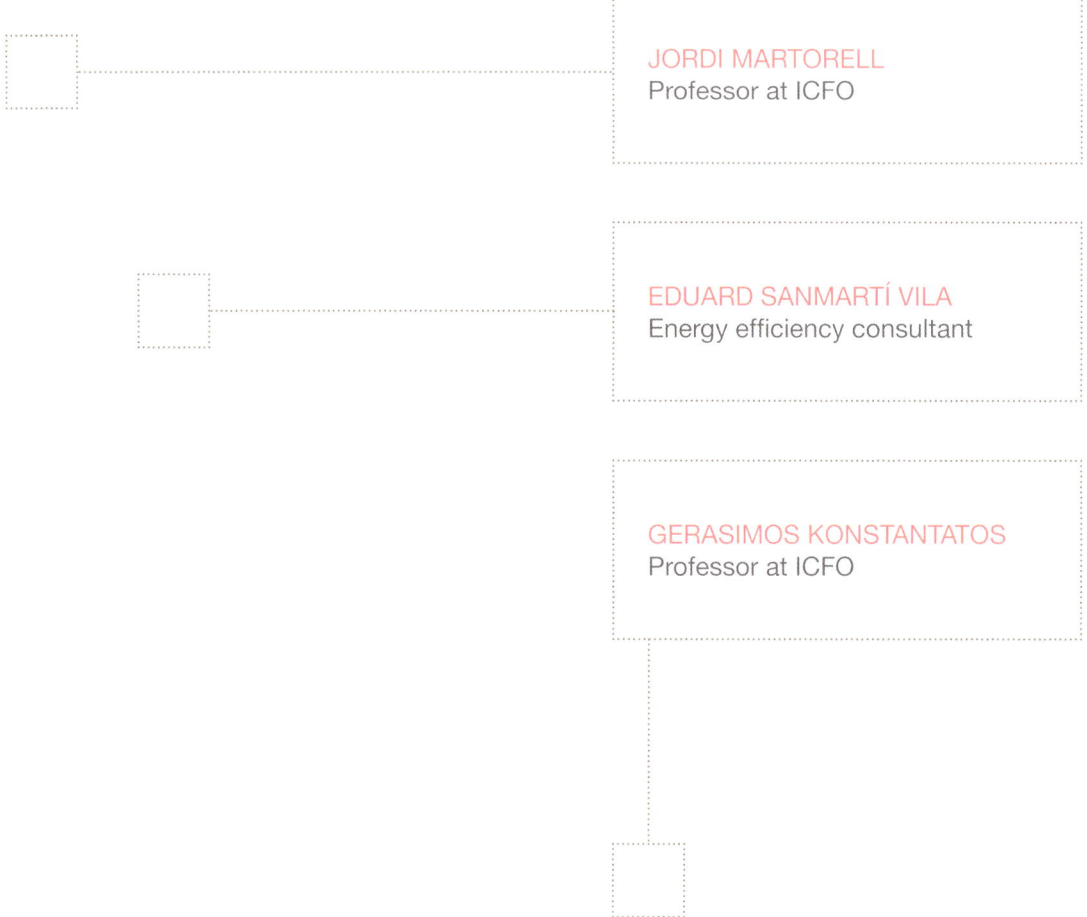

JORDI MARTORELL
Professor at ICFO

EDUARD SANMARTÍ VILA
Energy efficiency consultant

GERASIMOS KONSTANTATOS
Professor at ICFO

Glossary

INFRARED The portion of the electromagnetic spectrum extending from the red end of the visible region to the microwave range. The infrared range is usually divided into three regions: near-infrared, from 700 nm to 2 μm; medium-infrared, from 2 μm to 4 μm; and far-infrared, from 4 μm to 1 mm.

SILICON Chemical element in the carbon family, with symbol Si. After oxygen, it is the most abundant element of the Earth's crust. In its pure form and at ambient conditions, it is solid, hard, grey and has the same octahedral crystalline structure as the diamond form of carbon.

ULTRAVIOLET The portion of the electromagnetic spectrum extending from the violet end of the visible region to the X-ray range. The ultraviolet range is usually divided into three regions: near-ultraviolet, from 400 nm to 200 nm; far-ultraviolet, from 200 nm to 30 nm; and extreme-ultraviolet, from 30 nm to 10 nm.

VISIBLE REGION Portion of the electromagnetic spectrum to which the human eye is sensitive. It represents the peak of the Sun's spectrum. The visible wavelength's span extends from 700 nm (red) to 400 nm (violet).

Definitions adapted from
Oxford American English Dictionary
Encyclopaedia Britannica

Relevant reading

National Renewable Energies Laboratory
http://www.nrel.gov

Photovoltaic Geographical Information System (PVGIS)
http://re.jrc.ec.europa.eu/pvgis/

Fritzsche, H., Schwartz, B. B. (2008) *The Science and Technology of an American Genius: Stanford R. Ovshinsky*. World Scientific, Singapore

McEwan, I. (2010) *Solar*. Doubleday, New York

Tilley, R. J. D. (2011) *Colour and the Optical Properties of Materials: An Exploration of the Relationship between Light, the Optical Properties of Materials and Colour*. Wiley, Hoboken

Photo credits

p 108   Alessandro Volta's electric battery. Reproduced with permission of the *Musei civici di Como*

p 109   Solar Spectrum. © ICFO

LIGHTING

Life after sunset

Dedicated to the millions
of children who cannot read
a book at night.
We hope to see their number
constantly decrease over
the years.

LIGHTING

A steady light for reading in bed and keeping moving shadows at bay. Neons and LEDs fused with the city landscape. Candles on table-tops beside theaters. This is life after sunset. For many people growing up today, it is hard to grasp what it is like living without lightbulbs. The major impact of lighting technologies is directly addressed on their way and their quality of life. Besides enabling basic tasks, these technologies afford people the mere aesthetic pleasure of being surrounded by a particular play of light, one that they can design and control. In order to celebrate the ability to shape atmospheres with light, let's bring up the eloquence of three famous light-shaped scenes.

Gaslight (George Cukor, 1944) – life before electricity. London, late 1800s. Paula Alquist, a young woman recently married, starts to believe that she is losing her mind. Missing objects, fading lights, a number of minute details around her start changing, apparently unnoticed by others. After a bright night at the theater, Paula arrives home reassured. She's undressing when one of the gaslights in the room starts fading again. Her face turns from liveliness to suspicion and then panic. As the lights fade, and amidst the increasing shadows, she seems to bid farewell to the lights of her own sanity.

Rear Window (Alfred Hitchcock, 1955) – the lightbulb. The whole movie looks like a tribute to optics. The hero of the story, J. B. Jeffries, is a photojournalist, with the walls of his austere apartment in the Village covered with many of his award-winning photographs. His telephoto lens got him into the drama and the lights of his flash gave him a way out, blinding the murderer he helped to unveil. But it was a lightbulb (three, to be precise) that brought us here. While sleeping by the rear window, in almost complete darkness, a shadow approaches Jeff's face. The camera turns: a woman's face is illuminated by the backyard lights. When Jeff, jokingly, asks who she is, she stands up. She lights a table lamp: "Lisa" – pearl necklace, gracious hairstyle – "Carol" – another lamp, black fitted bodice, white tulle wrap – "Freemont" – the third lamp. The whole figure is revealed. Standing in her high heels, in a white skirt layered to mid-calf with chiffon and tulle, Lisa smiles triumphantly, surrounded by the silky glare of three lightbulbs.

Blade Runner (Ridley Scott, 1982) – city neons. The film is coming to an end with the final roof fight between Roy Batty, leader of the rebel Nexus-6 androids, and detective Deckard. The story takes place in a gloomy city of a post-crisis future, where the external features of androids – replicants – are almost indistinguishable from humans. The Sun is permanently covered. The foggy atmosphere is poorly illuminated by blue beacons from the top of some buildings, the headlights of gliding cars, and neon signs. At the verge of winning, Roy Batty spares Deckard's life and sits under a light rain to share with the human his last secret. A big TDK neon sign glows at his back. Its familiar blue, red, and green tubes could belong (and perhaps they did belong) to any city today. The sudden feeling of proximity prepares the audience for the final vindication: the replicant's deep feelings are indistinguishable from humans'.

Lighting sources

get
every now and then...

... RADICALLY MORE

ENABLING

IMMEDIATE

CONTROLLABLE

PORTABLE

SMALL

INTEGRABLE

AFFORDABLE

EFFICIENT

COLORFUL

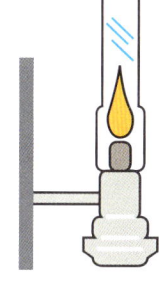

One challenge is finding
ways to make them
go all the way back to
natural.

LIGHTING
Life after sunset

In 1906, the inaugural address of the President of the Illuminating Engineering Society, L. B. Marks, concluded with the estimation that from the approximately $100 million spent by US consumers on electric light, 20% was due to an improper use of illumination, where he applied the term illumination to the use of light, in contraposition to the *production* of light. Historically, he remarked, all efforts had been concentrated on the *production* of light, while illumination, the use of light, was practically neglected, especially from the point of view of energy efficiency. As an "encouraging sign of the times," he identified a growing tendency of architecture and engineering to join efforts in dealing with problems involving both scientific and aesthetic aspects of illumination.

From that time to the present, things have evolved so that lighting companies, along with several energy-efficient improvements and alternatives to **incandescent** filament lamps, offer special personalized tutorials and advice to choose the optimal source for both comfort and energy efficiency.

Luminous efficacy of light sources is measured in **lumens** (lm) per watt (W). Main alternatives to classical incandescent filament lamps (approx. 15 lm/W) are **halogen** (approx. 25 lm/W), compact **fluorescent** (approx. 60 lm/W) and **LED** (approx. 80 lm/W). More recent sources include **OLED** panels (approx. 50 lm/W), mostly for interior design, and the inclusion of solar cell technology to eventually produce windows that act as transparent solar-energy collectors during daytime and lamps during the night.

Waste of light produced by a bad implementation of illumination, especially for outdoor lighting, not only has an economic impact. It also affects human health, nocturnal wildlife, seriously interferes with astronomical research, and extinguishes a starry sky in a night walk. The international campaign Globe at Night addresses the problem by inviting citizen-scientists to measure their night-sky brightness, submit their observations, and contribute to the public awareness of light pollution.

Life after
sunset

Scientific and technical advisors

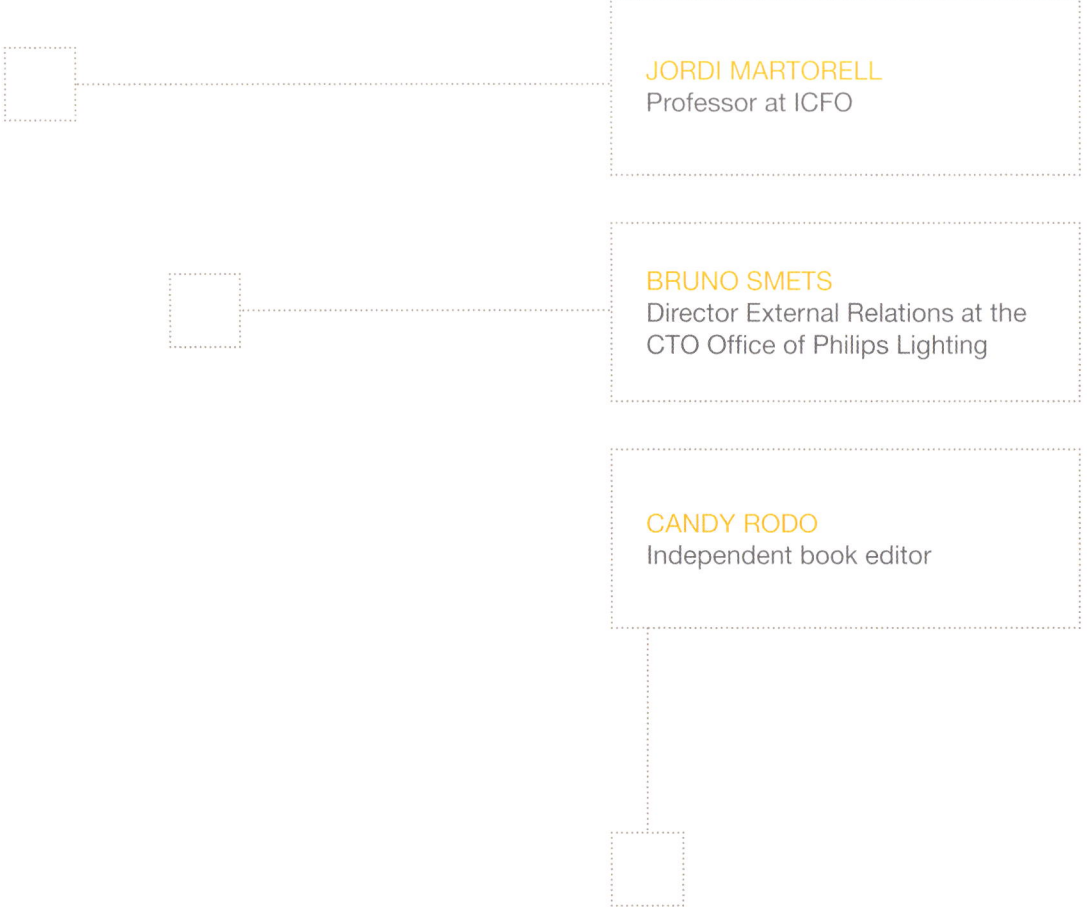

JORDI MARTORELL
Professor at ICFO

BRUNO SMETS
Director External Relations at the
CTO Office of Philips Lighting

CANDY RODO
Independent book editor

Glossary

....

FLUORESCENT LAMP Electric discharge lamp that produces light by the fluorescence of a phosphor coating. An electric arc passes through a mixture of an inert gas (argon or krypton) and a small amount of mercury. The mercury radiates ultraviolet energy that is transformed to visible light by the phosphor coating on the bulb.

HALOGEN A gas-filled tungsten filament incandescent lamp containing halogens or halogen compounds. The halogen combines with the tungsten evaporated from the hot filament to form a compound that is attracted back to the filament, thus extending the filament's life. The halogen regenerative cycle is also used to prevent blackening of the lamp envelope.

INCANDESCENT FILAMENT LAMPS A lamp in which light is produced by a filament heated to incandescence by an electric current. Normally, the filament is of coiled or coiled coil (doubly coiled) tungsten wire. However, it may be uncoiled wire, a flat strip, or of material other than tungsten.

LED Light-emitting diode. A semiconductor device which emits incoherent light when a voltage is applied. Efficiency continues to rise making an effective source of illumination for a huge number of applications.

LUMEN A measure of the total amount of visible light emitted by a source.

OLED Organic light-emitting diode. Light-emitting device composed of a matrix of diodes and organic light-emitting phosphors.

....

Definitions adapted from
Oxford American English Dictionary
Encyclopaedia Britannica

Relevant reading

Lighting the Way Ahead. McKinsey & Company. Available at www.mackinsey.com

Marks, L. B. (1906) Inaugural Address of President L. B. Marks. Available at http://www.ies.org

Jill, J. (2003) *Empires of Light: Edison, Tesla, Westinghouse, and the Race to Electrify the World*. Random House, New York

Alton, J. (1995) *Painting with Light*. University of California Press, Berkeley (first ed, 1949)

Salt, B. (1992) *Film Style and Technology: History and Analysis*. Starword, London

Photo credits

p 116 Earth at night, assembled from data by the Suomi NPP satellite © NASA Earth Observatory/NOAA NGDC

p 117 Sunset over Madrid from Plaza Callao © Ignacio Garrido

SENSING

Enhanced perception for everyday life

SENSING

SENSING

The different information channels that keep the human body in contact with the environment are traditionally grouped into five senses: taste, smell, touch, hearing, and sight. Taste and smell are mediated by chemical reactions at the nose and mouth; touch includes all the data gathered at diverse skin receptors: texture, temperature, pressure, humidity; hearing results from a mechanical process in the tympanum; and sight involves the interaction of light with the eye. However mediated through independent physical phenomena, human senses are not completely isolated from each other. They contribute to perception as an active team led by the brain to furnish every person with their own notion of what lies ahead.

In front of a coffee cup, senses complement, influence, and even compete with each other, to offer a full and personal impression of the coffee-drinking experience. Taste and smell lead to the flavor appreciation under a mutual influence. Touch and hearing may eventually supply conclusive details about the porcelain, complementing the visual inspection of the cup, which, in turn, exerts a non-negligible dominance on taste. Further connections, in every way more mysterious, are revealed by the synesthetic condition of some hypothetical individuals to whom the coffee's aroma might sound like an Elgar march.

Synesthesia is a peculiarity in perception that occurs when a stimulus over one sensorial channel triggers the perception of a different sense. It grounds the basis of some artificial approaches to sensing, since it implies the possibility of deliberately causing a sensorial impression without a direct stimulus. Certain sensorial impairments, like color-blindness, may be partially overcome by means of an implanted device that transduces the frequency of light corresponding to each color into a different sound frequency. As a result, colors, the notes of light, can be distinctively heard.

A beam of light has physical properties other than its color; like intensity, direction of propagation, or polarization. These properties may be affected in the interaction of light with different objects, revealing in the process diverse characteristics of such objects. The color of leaves indicates the presence of chlorophyll, the intensity of solar light indicates the thickness of the layer of clouds it crosses, changes in polarization orientate the flight for the specialized eyes of some insects. An extensive examination of the properties of light discloses rich amounts of information concerning several aspects of the objects it interacts with. To carry on such an examination, photonics technology is developing a wide range of sensors.

Since in most cases light travels in a straight line, at a constant speed, it makes an excellent distance sensor. Refraction and absorption of light on water droplets grounds humidity and rain detectors. The interaction of light with the chemical components of matter is used for quality control in the food industry or to monitor air pollution in cities. Optical fibers, integrated into structures, detect variations in temperature, torsion, or pressure. And so on and so forth.

Picture yourself in a car through a city, with tall smart buildings and endless displays. Thousands of sensors connected in networks embrace you into the city life. Your contribution to energy use, air quality, traffic flow, will be recorded and used for the welfare of all. Sensors' immediate diagnosis upgrade the capacity of machines, buildings, and virtually any object around to act as fine partners in more and more aspects of our everyday life.

In the most exaggerated, ludicrous future, the all-photonic supersensitive colleague standing overleaf would step out from your vehicle to share the full awareness coffee-break experience.

Subtle changes in light

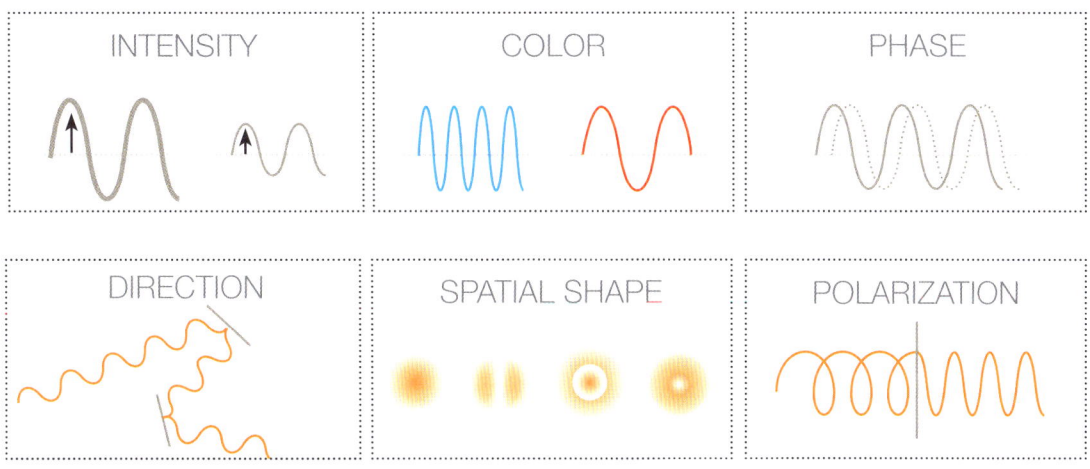

| INTENSITY | COLOR | PHASE |
| DIRECTION | SPATIAL SHAPE | POLARIZATION |

can be used to measure
with high precision

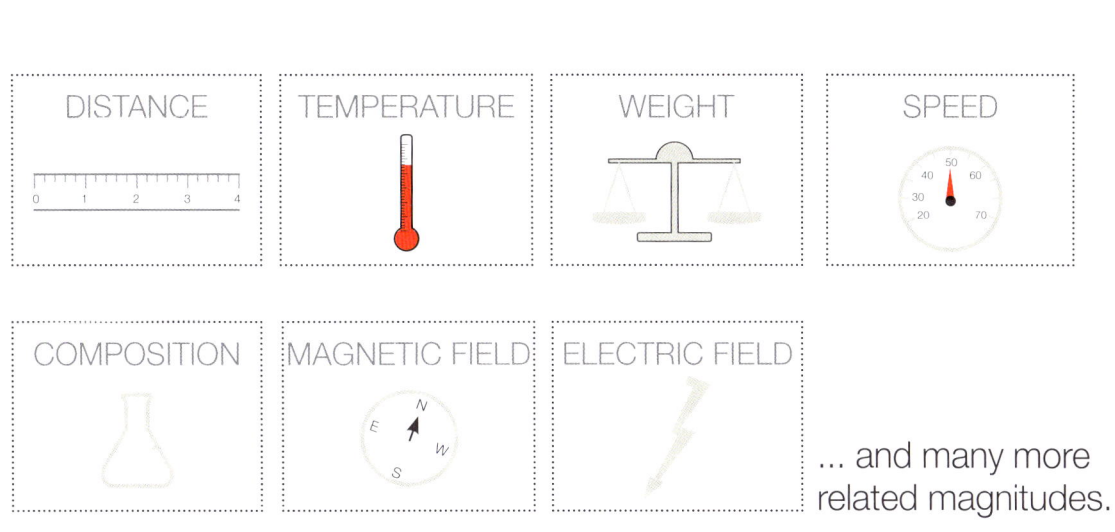

| DISTANCE | TEMPERATURE | WEIGHT | SPEED |
| COMPOSITION | MAGNETIC FIELD | ELECTRIC FIELD |

... and many more
related magnitudes.

The challenge is to build sensitive
receptors for such fine triggers.

Light is an electromagnetic wave. It can be described in terms of various parameters: direction of propagation, spatial shape, intensity, color (**wavelength**), **phase**, or **polarization**, the value of which can change by the interaction of light with matter. A light-based sensor is a device that determines features of a medium by assessing how light changes when it crosses it or reflects from it.

To give a simple image of how an optical-fiber sensor works, consider that light inside the fiber follows a zigzag course, confined along the walls by **total internal reflection**. The critical angle of the zigzag trajectory allowing total internal reflection – and therefore no losses – depends on the wavelength (the color) of the light and the **refractive index** of the core and the surrounding cladding. When a specific nano-patterning of the refractive index is created inside the fiber core, only certain colors are allowed to be transmitted while the others are reflected. External changes such as stretching, vibration, temperature, etc., cause a distortion on the refractive index pattern and, consequently, on the colors. By measuring such variations in color, it is possible to detect and quantify the external changes.

Light can also assess the composition of certain substances (in medical diagnosis, estimation of water and food quality, air monitoring, etc.) by means of several processes. **Spectroscopy** methods allow the chemical analysis of samples. Detection of specific substances with light is also possible with the aid of nanosized particles, which are able to concentrate the light in very small regions. Targeted substances would attach to the nanoparticle, affecting the color of the light there concentrated.

Free-space distance sensors work by illuminating an object and timing the delay in its return to the source. Small distances can be assessed by interferometry studying the phase difference between the emitted and received light. Changes in polarization can be used to measure the intensity and direction of magnetic fields by means of the **Faraday effect**.

VISCOSITY

COLOR

HUMIDITY

COMPOSITION

TEMPERATURE

PRESSURE

THICKNESS

QUALITY
CONTROL

WEIGHT

FLEXIBILITY

SOUND

SMELL

MAGNETIC
FIELD

STRAIN

ELECTRIC
FIELD

DISTANCE

VIBRATION

125

Scientific and technical advisors

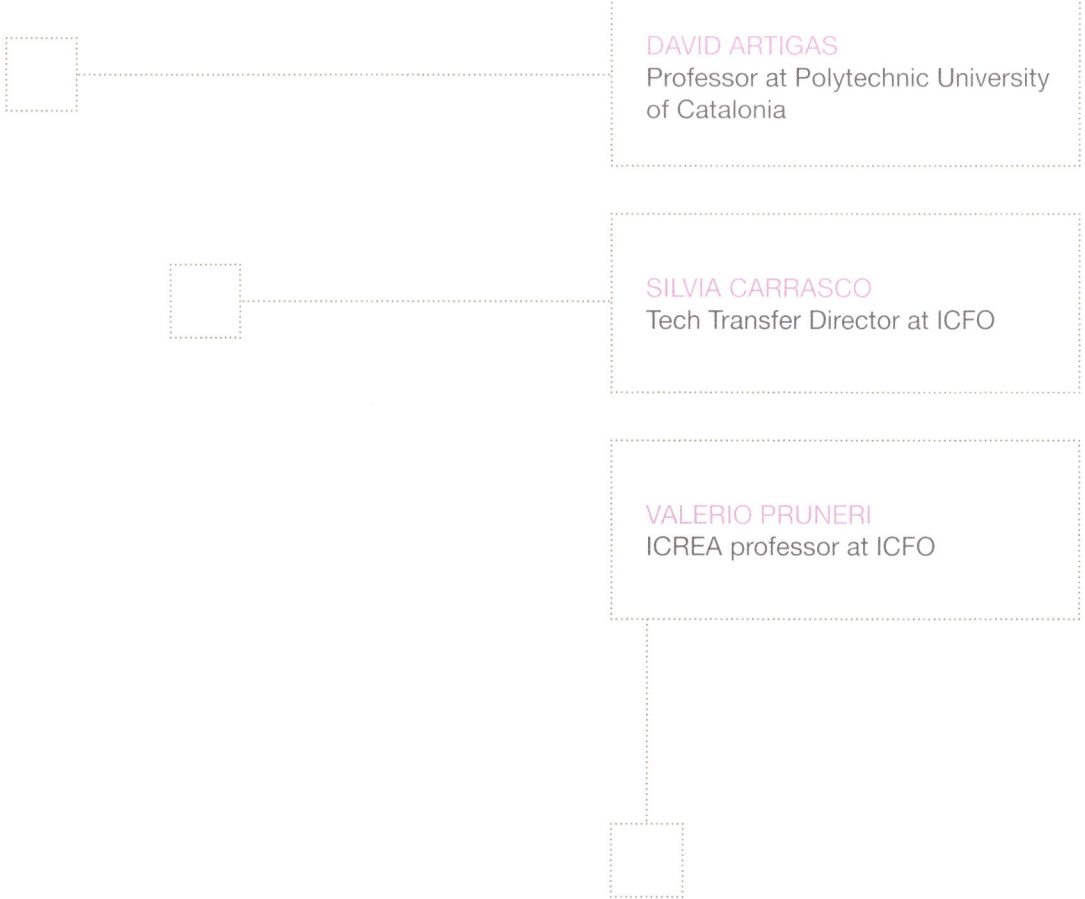

DAVID ARTIGAS
Professor at Polytechnic University
of Catalonia

SILVIA CARRASCO
Tech Transfer Director at ICFO

VALERIO PRUNERI
ICREA professor at ICFO

Glossary

....

FARADAY EFFECT The rotation of the plane of polarization of a light beam by a magnetic field.

PHASE The fraction of a wave that elapses relative to a fixed origin.

POLARIZATION In a beam of electromagnetic radiation, the polarization direction is the direction of the oscillation of the electric field of light. An individual photon has two possible states of polarization, in which information can be encoded.

REFRACTIVE INDEX Measure of the bending of a ray of light when passing from one medium into another.

SPECTROSCOPY Measurement of the spectra produced when matter interacts with or emits electromagnetic radiation.

TOTAL INTERNAL REFLECTION Complete reflection of a ray of light within a medium – such as water or glass – from the surrounding surfaces back into the medium. The phenomenon occurs if the angle of incidence is greater than a certain limiting angle, called the critical angle.

WAVELENGTH Electromagnetic energy is transmitted in the form of a sinusoidal wave. The wavelength is the physical distance covered by one cycle of this wave.

....

Definitions adapted from
Oxford American English Dictionary
Encyclopaedia Britannica

Relevant reading

Cutolo, A., Mignani, A. G., Tajani, A. (2014) *Photonics for Safety and Security*. World Scientific, Singapore

Righini, G. C., Tajani, A., Cutolo, A. (2009) *An Introduction to Optoelectronic Sensors*. World Scientific, Singapore

Haus, J. (2010) *Optical Sensors: Basics and Applications*. Wiley-VCH, Weinheim

Photo credits

p 124   LiDAR image of Whistling Straits Clubhouse, courtesy of James Young and Quantum Spatial

p 125   Steampunk robot, courtesy of Bjorn Hurri